Mechanical Engineering

Devang Saini

RANDOM PUBLICATIONS
NEW DELHI (INDIA)

Mechanical Engineering

ISBN 978-93-5111-424-6

Published in 2014 in India by

RANDOM PUBLICATIONS

4376-A/4B, Gali Murari Lal, Ansari Road
New Delhi-110 002
Phone : +91-11-43580356, +91-11-23289044
e-mail: randomexports@gmail.com, sales@randompublications.com, info@randompublications.com

Reprinted 2022

Type Setting by: Friends Media, Delhi-110089
Digitally Printed at: Replika Press Pvt. Ltd.

Preface

Applications of mechanical engineering are found in the records of many ancient and medieval societies throughout the globe. In ancient Greece, the works of Archimedes (287 BC–212 BC) deeply influenced mechanics in the Western tradition and Heron of Alexandria (c. 10–70 AD) created the first steam engine. In China, Zhang Heng (78–139 AD) improved a water clock and invented a seismometer, and Ma Jun(200–265 AD) invented a chariot with differential gears. The medieval Chinese horologist and engineer Su Song (1020–1101 AD) incorporated an escapement mechanism into his astronomical clock tower two centuries before any escapement can be found in clocks of medieval Europe, as well as the world's first known endless power-transmitting chain drive. During the years from 7th to 15th century, the era called the Islamic Golden Age, there were remarkable contributions from Muslim inventorsin the field of mechanical technology. Al-Jazari, who was one of them, wrote his famous Book of Knowledge of Ingenious Mechanical Devicesin 1206, and presented many mechanical designs. He is also considered to be the inventor of such mechanical devices which now form the very basic of mechanisms, such as the crankshaft and camshaft. Important breakthroughs in the foundations of mechanical engineering occurred in England during the 17th century when Sir Isaac Newtonboth formulated the three Newton's Laws of Motion and developed Calculus, the mathematical basis of physics. Newton was reluctant to publish his methods and laws for years, but he was finally persuaded to do so by his colleagues, such as Sir Edmund Halley, much to the benefit of all mankind. During the early 19th century in England, Germany and Scotland, the development of machine tools led mechanical engineering to develop as a separate field within engineering, providing manufacturing machines and the engines to power them. The first British professional society of mechanical engineers was formed in 1847 Institution of Mechanical Engineers, thirty years after the civil engineers formed the first such professional society Institution of Civil Engineers. On the European continent, Johann Von Zimmermann (1820–1901) founded the first factory for grinding machines in Chemnitz, Germany in 1848.

Mechanical engineering is a discipline of engineering that applies the principles of engineering, physics and materials science for analysis, design, manufacturing, and maintenance of mechanical systems. It is the branch of engineering that involves the production and usage of heat and mechanical power for the design, production, and operation of machines and tools. It is one of the oldest and broadest engineering disciplines. The engineering field requires an understanding of core concepts including mechanics, kinematics, thermodynamics, materials science, structural analysis, and electricity. Mechanical engineers use these core principles along with tools like computer-aided engineering, and product lifecycle management to design and analyse manufacturing plants, industrial equipment and machinery, heating and cooling systems, transport systems, aircraft, watercraft, robotics, medical devices, weapons, and others. Mechanical engineering emerged as a field during the industrial revolution in Europe in the 18th century; however, its development can be traced back several thousand years around the world. Mechanical engineering science emerged in the 19th century as a result of developments in the field of physics. The field has continually evolved to incorporate advancements in technology, and mechanical engineers today are pursuing developments in such fields as composites, mechatronics, and nanotechnology. Mechanical engineering overlaps with aerospace engineering, metallurgical engineering, civil engineering, electrical engineering, petroleum engineering, manufacturing engineering, chemical engineering, and other engineering disciplines to varying amounts. Mechanical engineers may also work in the field of Biomedical engineering, specifically with biomechanics, transport phenomena, biomechatronics, bionanotechnology and modelling of biological systems, like soft tissue mechanics.

This text is a one-stop resource for all researchers, teachers, and students involved in this crucial area of study.

I thank all members of my team who have helped in the preparation of the book. My special thanks go to "Random Publication" who have published the book.

—Devang Saini

Contents

1

System Models

An important concept in thermodynamics is the thermodynamic system, a precisely defined region of the universe under study. Everything in the universe except the system is known as the *surroundings*. A system is separated from the remainder of the universe by a *boundary* which may be notional or not, but which by convention delimits a finite volume. Exchanges of work, heat, or matter between the system and the surroundings take place across this boundary.

In practice, the boundary is simply an imaginary dotted line drawn around a volume when there is going to be a change in the internal energy of that volume. Anything that passes across the boundary that effects a change in the internal energy needs to be accounted for in the energy balance equation. The volume can be the region surrounding a single atom resonating energy, such as Max Planck defined in 1900; it can be a body of steam or air in a steam engine, such as Sadi Carnot defined in 1824; it can be the body of a tropical cyclone, such as Kerry Emanuel theorised in 1986 in the field of atmospheric thermodynamics; it could also be just one nuclide (i.e. a system of quarks) as hypothesised in quantum thermodynamics.

Boundaries are of four types: fixed, moveable, real, and imaginary. For example, in an engine, a fixed boundary means the piston is locked at its position; as such, a constant volume process occurs. In that same engine, a moveable boundary allows the piston to move in and out. For closed systems, boundaries are real while for open system boundaries are often imaginary.

In thermodynamic equilibrium, a system's properties are, by definition, unchanging in time. Systems in equilibrium are much simpler and easier to understand than systems which are not in equilibrium. Often, when analysing a thermodynamic process, it can be assumed that each intermediate state in the process is at equilibrium. This will also considerably simplify the situation. Thermodynamic processes which develop so slowly as to allow each intermediate step to be an equilibrium state are said to be reversible processes.

States and Processes

There are two fundamental kinds of entity in thermodynamics, states of a system, and processes of a system. This allows two fundamental approaches to thermodynamic reasoning, that in terms of states of a system, and that in terms of cyclic processes of a system.

The approach through states of a system requires a full account of the state of the system as well as a notion of process from one state to another of a system, but may require only a partial account of the state of the surroundings of the system or of other systems.

The notion of a cyclic process does not require a full account of the state of the system, but does require a full account of how the process occasions transfers of matter and energy between the system and its surroundings, which must include at least two heat reservoirs at different temperatures, one hotter than the other. In this approach, the notion of a properly numerical scale of temperature is a presupposition of thermodynamics, not a notion constructed by or derived from it.

Thermodynamic State Variables

When a system is at thermodynamic equilibrium under a given set of conditions of its surroundings, it is said to be in a definite thermodynamic state, which is fully described by its state variables.

Thermodynamic state variables are of two kinds, extensive and intensive. Examples of extensive thermodynamic variables are total mass and total volume. Examples of intensive thermodynamic variables are temperature, pressure, and chemical concentration; intensive thermodynamic variables are defined at each spatial point and each instant of time in a system. Physical macroscopic variables can be mechanical or thermal. Temperature is a thermal variable; according to Guggenheim, "the most important conception in thermodynamics is temperature."

If a system is in thermodynamic equilibrium and is not subject to an externally imposed force field, such as gravity, electricity, or magnetism, then (subject to a proviso stated in the following sentence) it is homogeneous, that is say, spatially uniform in all respects. There is a proviso here; a system in thermodynamic equilibrium can be inhomogeneous in the following respect: it can consist of several so-called 'phases', each homogeneous in itself, in immediate contiguity with other phases of the system, but distinguishable by their having various respectively different physical characters; a mixture of different chemical species is considered homogeneous for this purpose if it is physically homogeneous. For example, a vessel can contain a system consisting of water vapour overlying liquid water; then there is a vapour phase and a liquid phase, each homogeneous in itself, but still in thermodynamic equilibrium with the other phase. For the immediately present account, systems with multiple phases are not considered, though for many thermodynamic questions, multiphase systems are important.

In a sense, a homogeneous system can be regarded as spatially zero-dimensional, because it has no spatial variation.

If a system in thermodynamic equilibrium is homogeneous, then its state can be described by a number of intensive variables and extensive variables.

Intensive variables are defined by the property that if any number of systems, each in its own separate homogeneous thermodynamic equilibrium state, all with the same respective values of all of their intensive variables, regardless of the values of their extensive variables, are laid contiguously with no partition between them, so as to form a new system, then the values of the intensive variables of the new system are the same as those of the separate constituent systems. Such a composite system is in a homogeneous thermodynamic equilibrium. Examples of intensive variables are temperature, chemical concentration, pressure, density of mass, density of internal energy, and, when it can be properly defined, density of entropy.

Extensive variables are defined by the property that if any number of systems, regardless of their possible separate thermodynamic equilibrium or non-equilibrium states or intensive variables, are laid side by side with no partition between them so as to form a new system, then the values of the extensive variables of the new system are the sums of the values of the respective extensive variables of the individual separate constituent systems. Obviously, there is no reason to expect such a composite system to be in in a homogeneous thermodynamic equilibrium. Examples of extensive variables are mass, volume, and internal energy. They depend on the total quantity of mass in the system.

Though, when it can be properly defined, density of entropy is an intensive variable, entropy itself does not fit into this classification of state variables. The reason is that entropy is a property of a system as a whole, and not necessarily related simply to its constituents separately. It is true that for any number of systems each in its own separate homogeneous thermodynamic equilibrium, all with the same values of intensive variables, removal of the partitions between the separate systems results in a composite homogeneous system in thermodynamic equilibrium, with all the values of its intensive variables the same as those of the constituent systems, and it is reservedly or conditionally true that the entropy of such a restrictively defined composite system is the sum of the entropies of the constituent systems.

But if the constituent systems do not satisfy these restrictive conditions, the entropy of a composite system cannot be expected to be the sum of the entropies of the constituent systems, because the entropy is a property of the composite system as a whole. Therefore, though under these restrictive reservations, entropy satisfies some requirements for extensivity defined just above, entropy in general does not fit the above definition of an extensive variable.

Being neither an intensive variable nor an extensive variable according to the above definition, entropy is thus a stand-out variable, because it is a state variable of a system as a whole. A non-equilibrium system can have a very inhomogeneous dynamical structure. This is one reason for distinguishing the study of equilibrium thermodynamics from the study of non-equilibrium thermodynamics.

The physical reason for the existence of extensive variables is the time-invariance of volume in a given inertial reference frame, and the strictly local conservation of mass, momentum, angular momentum, and energy. As noted by Gibbs, entropy is unlike energy and mass, because it is not locally conserved. The stand-out quantity entropy is never conserved in real physical processes; all real physical processes are irreversible.

The motion of planets seems reversible on a short time scale (millions of years), but their motion, according to Newton's laws, is mathematically an example of deterministic chaos. Eventually a planet will suffer an unpredictable collision with an object from its surroundings, outer space in this case, and consequently its future course will be radically unpredictable. Theoretically this can be expressed by saying that every natural process dissipates some information from the predictable part of its activity into the unpredictable part. The predictable part is expressed in the generalised mechanical variables, and the unpredictable part in heat.

There are other state variables which can be regarded as conditionally 'extensive' subject to reservation as above, but not extensive as defined above. Examples are the Gibbs free energy, the Helmholtz free energy, and the enthalpy. Consequently, just because for some systems under particular conditions of their surroundings such state variables are conditionally conjugate to intensive variables, such conjugacy does not make such state variables extensive as defined above. This is another reason for distinguishing the study of equilibrium thermodynamics from the study of non-equilibrium thermodynamics. In another way of thinking, this explains why heat is to be regarded as a quantity that refers to a process and not to a state of a system.

Equation of State

The properties of a system can under some conditions be described by an equation of state which specifies the relationship between state variables. The equation of state expresses constitutive properties of the material of the system. It must comply with some thermodynamic constraints, but cannot be derived from the general principles of thermodynamics alone.

Thermodynamic Processes

A thermodynamic process is defined by changes of state internal to the system of interest, combined with transfers of matter and energy to and from the surroundings of the system or to and from other systems. A system is demarcated from its surroundings or from other systems by partitions which may more or less separate them, and may move as a piston to change the volume of the system and thus transfer work.

Dependent and Independent Variables for a Process

A process is described by changes in values of state variables of systems or by quantities of exchange of matter and energy between systems and surroundings. The change must be specified in terms of prescribed variables. The choice of which variables are to be used is made in advance of consideration of the course of the process, and cannot be changed. Certain of the variables chosen in advance are called the independent variables. From changes in independent variables may be derived changes in other variables called dependent variables. For example a process may occur at constant pressure with pressure prescribed as an independent variable, and temperature changed as another independent variable, and then changes in volume are considered as dependent. Careful attention to this principle is necessary in thermodynamics.

Changes of State of a System

In the approach through states of the system, a process can be described in two main ways.

In one way, the system is considered to be connected to the surroundings by some kind of more or less separating partition, and allowed to reach equilibrium with the surroundings with that partition in place. Then, while the separative character of the partition is kept unchanged, the conditions of the surroundings are changed, and exert their influence on the system again through the separating partition, or the partition is moved so as to change the volume of the system; and a new equilibrium is reached. For example, a system is allowed to reach equilibrium with a heat bath at one temperature; then the temperature of the heat bath is changed and the system is allowed to reach a new equilibrium; if the partition allows conduction of heat, the new equilibrium will be different from the old equilibrium.

In the other way, several systems are connected to one another by various kinds of more or less separating partitions, and to reach equilibrium with each other, with those partitions in place. In this way, one may speak of a 'compound system'. Then one or more partitions is removed or changed in its separative properties or moved, and a new equilibrium is reached. The Joule-Thomson experiment is an example of this; a tube of gas is separated from another tube by a porous partition; the volume available in each of the tubes is determined by respective pistons; equilibrium is established with an initial set of volumes; the volumes are changed and a new equilibrium is established. Another example is in separation and mixing of gases, with use of chemically semi-permeable membranes.

Cyclic Processes

A cyclic process is a process that can be repeated indefinitely often without changing the final state of the system in which the process occurs. The only traces of the effects of a cyclic process are to be found in the surroundings of the system or in other systems. This is the kind of process that concerned early thermodynamicists such as Carnot, and in terms of which Kelvin defined absolute temperature, before the use of the quantity of entropy by Rankine and its clear identification by Clausius. For some systems, for example with some plastic working substances, cyclic processes are practically nearly unfeasible because the working substance undergoes practically irreversible changes. This is why mechanical devices are lubricated with oil and one of the reasons why electrical devices are often useful.

A cyclic process of a system requires in its surroundings at least two heat reservoirs at different temperatures, one at a higher temperature that supplies heat to the system, the other at a lower temperature that accepts heat from the system. The early work on thermodynamics tended to use the cyclic process approach, because it was interested in machines which would convert some of the heat from the surroundings into mechanical power delivered to the surroundings, without too much concern about the internal workings of the machine. Such a machine, while receiving an amount of heat from a higher temperature reservoir, always needs a lower temperature reservoir that accepts some lesser amount of heat, the difference in amounts of heat being converted to work. Later, the internal workings of a system became of interest, and they are described by the states of the system. Nowadays, instead of arguing in terms of cyclic processes, some writers are inclined to derive the concept of absolute temperature from the concept of entropy, a variable of state.

Commonly Considered Thermodynamic Processes

It is often convenient to study a thermodynamic process in which a single variables, such as temperature, pressure, or volume, etc., is held fixed. Furthermore, it is useful to group these processes into pairs, in which each variable held constant is one member of a conjugate pair.

Several commonly studied thermodynamic processes are:

Isobaric Process

An isobaric process is a thermodynamic process in which the pressure stays constant. The term derives from the Greek *isos,* (equal), and *barus,* (heavy). The heat transferred to the system does work but also changes the internal energy of the system:

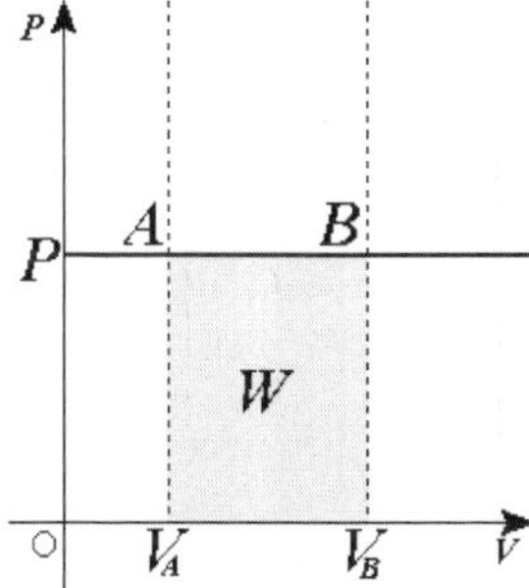

Figure: *The grey area represents the work done*

$$Q = \Delta U + W$$

According to the first law of thermodynamics, where W is work done by the system, U is internal energy, and Q is heat. Pressure-volume work by the closed system is defined as:

$$W = \int p dV$$

where Δ means change over the whole process, whereas d denotes a differential. Since pressure is constant, this means that

$$W = p\Delta V \ .$$

Applying the ideal gas law, this becomes

$$W = nR\Delta T$$

assuming that the quantity of gas stays constant, e.g., there is no phase change during a chemical reaction. According to the equipartition theorem, the change in internal energy is related to the temperature of the system by

$$\Delta U = nc_V \Delta T \ ,$$

where c_V is specific heat at a constant volume.

Substituting the last two equations into the first equation produces:

$$\begin{aligned} Q &= nc_V \Delta T + nR\Delta T \\ &= n(c_V + R)\Delta T \\ &= nc_P \Delta T \end{aligned}$$

where c_P is specific heat at a constant pressure.

Specific Heat Capacity

To find the molar specific heat capacity of the gas involved, the following equations apply for any general gas that is calorically perfect. The property γ is either called the adiabatic index or the heat capacity ratio. Some published sources might use k instead of γ.

Molar isochoric specific heat:

$$c_V = \frac{R}{\gamma - 1} \ .$$

Molar isobaric specific heat:

$$c_p = \frac{\gamma R}{\gamma - 1} \ .$$

The values for γ are $\gamma = 1.4$ for diatomic gasses like air and its major components, and $\gamma = \frac{5}{3}$ for monatomic gasses like the noble gasses. The formulas for specific heats would reduce in these special cases:

Monatomic: $c_V = \frac{3R}{2}$

and $c_P = \frac{5R}{2}$

Diatomic: $c_V = \frac{5R}{2}$

and $c_P = \frac{7R}{2}$

If the process moves towards the right, then it is an expansion. If the process moves towards the left, then it is a compression.

Sign Convention for Work

The motivation for the specific sign conventions of thermodynamics comes from early development of heat engines. When designing a heat engine, the goal is to have the system produce and deliver work output. The source of energy in a heat engine, is a heat input.

If the volume compresses (delta V = final volume - initial volume < 0), then W < 0. That is, during isobaric compression the gas does negative work, or the environment does positive work. Restated, the environment does positive work on the gas.

If the volume expands (delta V = final volume - initial volume > 0), then W > 0. That is, during isobaric expansion the gas does positive work, or equivalently, the environment does negative work. Restated, the gas does positive work on the environment.

If heat is added to the system, then Q > 0. That is, during isobaric expansion/heating, positive heat is added to the gas, or equivalently, the environment receives negative heat. Restated, the gas receives positive heat from the environment. If the system rejects heat, then Q < 0. That is, during isobaric compression/cooling, negative heat is added to the gas, or equivalently, the environment receives positive heat. Restated, the environment receives positive heat from the gas.

Isochoric Process

An isochoric process, also called a constant-volume process, an isovolumetric process, or an isometric process, is a thermodynamic process during which the volume of the closed system undergoing such a process remains constant. An isochoric process is exemplified by the heating or the cooling of the contents of a sealed, inelastic container: The thermodynamic process is the addition or removal of heat; the isolation of the contents of the container establishes the closed system; and the inability of the container to deform imposes the constant-volume condition.

Formalism

An isochoric thermodynamic process is characterised by constant volume, i.e. $\Delta V = 0$. The process does no pressure-volume work, since such work is defined by

$$\Delta W = P\Delta V,$$

where P is pressure. The sign convention is such that positive work is performed by the system on the environment.

For a reversible process, the first law of thermodynamics gives the change in the system's internal energy:

$$dU = dQ - dW$$

Replacing work with a change in volume gives

$$dU = dQ - PdV$$

Since the process is isochoric, $dV = 0$, the previous equation now gives

$$dU = dQ$$

Using the definition of specific heat capacity at constant volume,

$$C_v = dU / dT,$$

$$dQ = nC_v dT$$

Integrating both sides yields

$$Q = n\int_a^b C\,dT.$$

Where C is the specific heat capacity at constant volume, a is initial temperature and b is final temperature. We conclude with:

$$Q = nC_v\Delta T$$

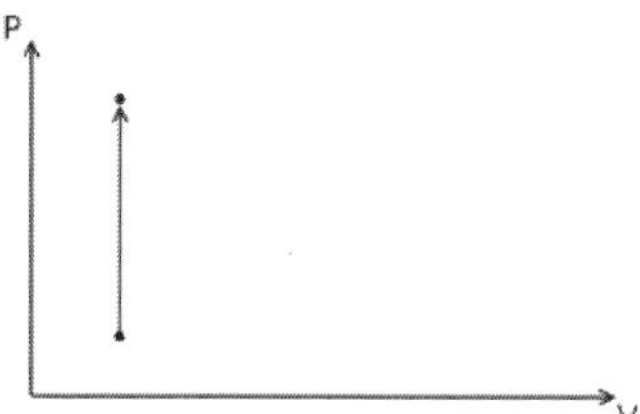

Figure: *Isochoric Process in the Pressure volume diagram.*

In this diagram, pressure increases, but volume remains constant.

On a pressure volume diagram, an isochoric process appears as a straight vertical line. Its thermodynamic conjugate, an isobaric process would appear as a straight horizontal line.

Ideal Gas

If an ideal gas is used in an isochoric process, and the quantity of gas stays constant, then the increase in energy is proportional to an increase in temperature and pressure. Take for example a gas heated in a rigid container: the pressure and temperature of the gas will increase, but the volume will remain the same.

Ideal Otto Cycle

The ideal Otto cycle is an example of an isochoric process when it is assumed that the burning of the gasoline-air mixture in an internal combustion engine car is instantaneous. There is an increase in the temperature and the pressure of the gas inside the cylinder while the volume remains the same.

Etymology

The noun isochor and the adjective isochoric are derived from the Greek words *isos* meaning "equal", and ÷*þρα* (*chora*) meaning "space."

Isothermal Process

An isothermal process is a change of a system, in which the temperature remains constant: $\Delta T = 0$. This typically occurs when a system is in contact with an outside thermal reservoir (heat bath), and the change occurs slowly enough to allow the system to continually adjust to the temperature of the reservoir through heat exchange. In contrast, an *adiabatic process* is where a system exchanges no heat with its surroundings ($Q = 0$). In other words, in an isothermal process, the value $\Delta T \neq 0$ but $Q \neq 0$, while in an adiabatic process, $\Delta T \neq 0$ but $Q = 0$.

Details for an Ideal Gas

For the special case of a gas to which Boyle's law applies, the product pV is a constant if the gas is kept at isothermal conditions. The value of the constant is nRT, where n is the number of moles of gas present and R is the ideal gas constant. In other words, the ideal gas law $pV = nRT$ applies. This means that

$$p = \frac{nRT}{V} = \frac{\text{constant}}{V}$$

holds. The family of curves generated by this equation is shown in the graph presented at the bottom right-hand of the page. Each curve is called an isotherm. Such graphs are termed indicator diagrams and were first used by James Watt and others to monitor the efficiency of engines. The temperature corresponding to each curve in the figure increases from the lower left to the upper right.

Calculation of Work

In thermodynamics, the work involved when a gas changes from state A to state B is simply

$$W_{A\to B} = \int_{V_A}^{V_B} p dV$$

For an isothermal, reversible process, this integral equals the area under the relevant pressure-volume isotherm, and is indicated in blue in the figure for an ideal gas. Again, $p = nRT / V$ applies and with T being constant (as this is an isothermal process), we have:

$$W_{A\to B} = \int_{V_A}^{V_B} p dV = \int_{V_A}^{V_B} \frac{nRT}{V} dV = nRT \int_{V_A}^{V_B} \frac{1}{V} dV = nRT \ln \frac{V_B}{V_A}$$

By convention, work is defined as the work the system does on its environment. If, for example, the system expands by a piston moving in the direction of force applied by the internal pressure of a gas, then the work is counted as positive, and as this work is done by using internal energy of the system, the result is that the internal energy decreases. Conversely, if the environment does work on the system so that its internal energy increases, the work is counted as negative.

It is also worth noting that, for many systems, if the temperature is held constant, the internal energy of the system also is constant, and so $\Delta U = 0$. From First Law of Thermodynamics, $Q = \Delta U + W$, so it follows that $Q = W$ for this same isothermal process.

Applications

Isothermal processes can occur in any kind of system, including highly-structured machines, and even living cells. Various parts of the cycles of some heat engines are carried out isothermally and may be approximated by a Carnot cycle. Phase changes, such as melting or evaporation, are also isothermal processes.

In Isothermal non-flow Process, the work done by compressing the perfect gas (Pure Substance) is a negative work, as work is done on the system, as result of compression, the volume will decrease, and temperature will try to increase. To maintain the temperature at constant value (as the process is isothermal) heat energy has to leave the system and enter the environment. The amount of energy entering the environment is equal to the work done (by compressing the perfect gas) because internal energy does not change. The thermodynamic sign convention is that heat entering the environment is also negative. Thence Q = W.

In equation of work, the term nRT can be replaced by PV of any state. The product of pressure and volume is in fact, 'Moving Boundary Work'; the systems boundaries are compressed. For Expansion the same theory is applied.

As per Joule's Law, Internal energy is the function of absolute temperature. In isothermal process the temperature is constant. Hence, the internal energy is constant. And net change in internal energy is ZERO.

Adiabatic Process

In thermodynamics, an adiabatic process or an isocaloric process is a thermodynamic process in which the net heat transfer to or from the working fluid is zero. Such a process can occur if the container of the system has thermally-insulated walls or the process happens in an extremely short time, so that there is no opportunity for significant heat exchange. The term "adiabatic" literally means impassable, coming from the Greek roots not, through, and to pass; this etymology corresponds here to an absence of heat transfer. Conversely, a process that involves heat transfer (addition or loss of heat to the surroundings) is generally called diabatic. Although the terms adiabatic and isocaloric can often be interchanged, adiabatic processes may be considered a subset of isocaloric processes; the remaining complement subset of isocaloric processes being processes where net heat transfer does not diverge regionally such as in an idealised case with mediums of infinite thermal conductivity or non-existent thermal capacity.

In an adiabatic irreversible process, $dQ = 0$ is not equal to TdS ($TdS > 0$). $dQ = TdS = 0$ holds for reversible processes only. For example, an adiabatic boundary is a boundary that is impermeable to heat transfer and the system is said to be adiabatically (or thermally) insulated; an insulated wall approximates an adiabatic boundary. Another example is the adiabatic flame temperature, which is the temperature that would be achieved by a flame in the absence of heat loss to the surroundings. An adiabatic process that is reversible is also called an isentropic process. Additionally, an adiabatic process that is irreversible and extracts no work is in an isenthalpic process, such as viscous drag, progressing towards a non-negative change in entropy.

One opposite extreme—allowing heat transfer with the surroundings, causing the temperature to remain constant—is known as an isothermal process. Since temperature is thermodynamically conjugate to entropy, the isothermal process is conjugate to the adiabatic process for reversible transformations.

A transformation of a thermodynamic system can be considered adiabatic when it is quick enough that no significant heat is transferred between the system and the outside. At the opposite extreme, a transformation of a thermodynamic system can be considered isothermal if it is slow enough so that the system's temperature remains constant by heat exchange with the outside.

Adiabatic Heating and Cooling

Adiabatic changes in temperature occur due to changes in pressure of a gas while not adding or subtracting any heat. In contrast, free expansion is an isothermal process for an ideal gas.

Adiabatic heating occurs when the pressure of a gas is increased from work done on it by its surroundings, e.g. a piston. Diesel engines rely on adiabatic heating during their compression stroke to elevate the temperature sufficiently to ignite the fuel.

Adiabatic heating also occurs in the Earth's atmosphere when an air mass descends, for example, in a katabatic wind or Foehn wind flowing downhill. When a parcel of air descends, the pressure on the parcel increases. Due to this increase in pressure, the parcel's volume decreases and its temperature increases, thus increasing the internal energy.

Adiabatic cooling occurs when the pressure of a substance is decreased as it does work on its surroundings. Adiabatic cooling does not have to involve a fluid. One technique used to reach very low temperatures (thousandths and even millionths of a degree above absolute zero) is adiabatic demagnetisation, where the change in magnetic field on a magnetic material is used to provide adiabatic cooling. Adiabatic cooling occurs in the Earth's atmosphere with orographic lifting and lee waves, and this can form pileus or lenticular clouds if the air is cooled below the dew point. Also, the contents of an expanding universe (to first order) can be described as an adiabatically cooling fluid. When the pressure applied on a parcel of air decreases, the air in the parcel is allowed to expand; as the volume increases, the temperature falls and internal energy decreases.

Rising magma also undergoes adiabatic cooling before eruption.

Such temperature changes can be quantified using the ideal gas law, or the hydrostatic equation for atmospheric processes.

No process is truly adiabatic. Many processes are close to adiabatic and can be easily approximated by using an adiabatic assumption, but there is always some heat loss; as no perfect insulators exist.

Isentropic Process

In thermodynamics, an isentropic process or isoentropic process "disorder" is one in which for purposes of engineering analysis and calculation, one may assume that the process takes place from initiation to completion without an increase or decrease in the entropy of the system, i.e., the entropy of the system remains constant. It can be proven that any reversible adiabatic process is an isentropic process.

Background

The Second law of thermodynamics states that,

$$\delta Q \leq T dS$$

where δQ is the amount of energy the system gains by heating, T is the temperature of the system, and dS is the change in entropy. The equal sign will hold for a reversible process. For a reversible isentropic process, there is no transfer of heat energy and therefore the process is also adiabatic. For an irreversible process, the entropy will increase. Hence removal of heat from the system (cooling) is necessary to maintain a constant internal entropy for an irreversible process in order to make it isentropic. Thus an irreversible isentropic process is not adiabatic.

For reversible processes, an isentropic transformation is carried out by thermally "insulating" the system from its surroundings. Temperature is the thermodynamic conjugate variable to entropy, thus the conjugate process would be an isothermal process in which the system is thermally "connected" to a constant-temperature heat bath.

Isentropic Flow

An isentropic flow is a flow that is both adiabatic and reversible. That is, no heat is added to the flow, and no energy transformations occur due to friction or dissipative effects. For an isentropic flow of a perfect gas, several relations can be derived to define the pressure, density and temperature along a streamline.

Note that energy can be exchanged with the flow in an isentropic transformation, as long as it doesn't happen as heat exchange. An example of such an exchange would be an isentropic expansion or compression, with entail work done on or by the flow.

Isenthalpic Process

An isenthalpic process or isoenthalpic process is a process that proceeds without any change in enthalpy, H; or specific enthalpy, h.

In a steady-state, steady-flow process, significant changes in pressure and temperature can occur to the fluid and yet the process will be isenthalpic if there is no transfer of heat to or from the surroundings, no work done on or by the surroundings, and no change in the kinetic energy of the fluid. (If a steady-state, steady-flow process is analysed using a control volume everything outside the control volume is considered to be the surroundings.)

The throttling process is a good example of an isenthalpic process. Consider the lifting of a relief valve or safety valve on a pressure vessel. The specific enthalpy of the fluid inside the pressure vessel is the same as the specific enthalpy of the fluid as it escapes from the valve. With a knowledge of the specific enthalpy of the fluid, and the pressure outside the pressure vessel, it is possible to determine the temperature and speed of the escaping fluid.

In an isenthalpic process:

- $h_1 = h_2$
- $dh = 0$

Isenthalpic processes on an ideal gas follow isotherms since $dh = 0 = c_p dT$.

2

Applications for Technical Drawing

Architecture

The art and design that goes into making buildings is known as architecture. To communicate all aspects of the design, detailed drawings are used. In this field, the term plan is often used when referring to the full section view of these drawings. Architectural drawings describe and document an architect's design.

Engineering

Engineering can be a very broad term. It stems from the Latin *ingenerare*, meaning "to create". Because this could apply to everything that humans create, it is given a narrower definition in the context of technical drawing. Engineering drawings generally deal with mechanical engineered items, such as manufactured parts and equipment.

Engineering drawings are usually created in accordance with standardised conventions for layout, nomenclature, interpretation, appearance size, etc. Its purpose is to accurately and unambiguously capture all the geometric features of a product or a component. The end goal of an engineering drawing is to convey all the required information that will allow a manufacturer to produce that component.

Related Fields

Technical Illustrations

Technical illustration is the use of illustration to visually communicate information of a technical nature. Technical illustrations can be component technical drawings or diagrams. The aim of technical illustration is "to generate expressive images that effectively convey certain information via the visual channel to the human observer".

The main purpose of technical illustration is to describe or explain these items to a more or less non-technical audience. The visual image should be accurate in terms of dimensions and proportions, and should provide "an overall impression of what an object is or does, to enhance the viewer's interest and understanding".

According to Viola (2005) "illustrative techniques are often designed in a way that even a person with no technical understanding clearly understands the piece of art. The use of varying line widths to emphasise mass, proximity, and scale helped to make a simple line drawing more understandable to the lay person. Cross hatching, stippling, and other low abstraction techniques gave greater depth and dimension to the subject matter".

A cutaway drawing is a technical illustration, in which surface elements a three-dimensional model are selectively removed, to make internal features visible, but without sacrificing the outer context entirely.

The purpose of a cutaway drawing is to "allow the viewer to have a look into an otherwise solid opaque object. Instead of letting the inner object shine through the surrounding surface, parts of outside object are simply removed. This produces a visual appearance as if someone had cutout a piece of the object or sliced it into parts. Cutaway illustrations avoid ambiguities with respect to spatial ordering, provide a sharp contrast between foreground and background objects, and facilitate a good understanding of spatial ordering".

Types of Technical Drawings

The two types of technical drawings are based on graphical projection. This is used to create an image of a three-dimensional object onto a two-dimensional surface.

Two-Dimensional Representation

Two-dimensional representation uses orthographic projection to create an image where only two of the three dimensions of the object are seen.

Three-Dimensional Representation

In three-dimensional representation, also referred to as pictorials, all three dimensions, of the three dimensions of an object, are visible.

Multiview is a type of orthographic projection. There are two conventions for using multiview, first-angle and third-angle. In both cases, the front or main side of the object is the same. First-angle is drawing the object sides based on where they land. Example, looking at the front side, rotate the object 90 degrees to the right. What is seen will be drawn to the right of the front side. Third-angle is drawing the object sides based on where they are. Example, looking at the front side, rotate the object 90 degrees to the right. What is seen is actually the left side of the object and will be drawn to the left of the front side

Section

While multiview relates to external surfaces of an object, section views show an imaginary plane cut through an object. This is often useful to show voids in an object.

Auxiliary

Auxiliary views utilise an additional projection plane other than the common planes in a multiview. Since the features of an object need to show the true shape and size of the object, the projection plane must be parallel to the object surface. Therefore, any surface that is not in line with the three major axis needs its own projection plane to show the features correctly.

Pattern

Patterns, sometimes called developments, show the size and shape of a flat piece of material needed for later bending or folding into a three dimensional shape.

Exploded

An exploded view drawing is a technical drawing of an object that shows the relationship or order of assembly of the various parts. It shows the components of an object slightly separated by distance, or suspended in surrounding space in the case of a three-dimensional exploded diagram. An object is represented as if there had been a small controlled explosion emanating from the middle of the object, causing the object's parts to be separated relative distances away from their original locations.

An exploded view drawing (EVD) can show the intended assembly of mechanical or other parts. In mechanical systems usually the component closest to the center is assembled first, or is the main part in which the other parts get assembled. This drawing can also help to represent disassembly of parts, where the parts on the outside normally get removed first.

Standards and Conventions

Basic Drafting Paper Sizes: There have been many standard sizes of paper at different times and in different countries, but today most of the world uses the international standard (A4 and its siblings). North America uses its own sizes.

Patents

The applicant for a patent will be required by law to furnish a drawing of the invention whenever the nature of the case requires a drawing to understand the invention. This drawing must be filed with the application. This includes practically all inventions except compositions of matter or processes, but a drawing may also be useful in the case of many processes.

The drawing must show every feature of the invention specified in the claims, and is required by the patent office rules to be in a particular form. The Office specifies the size of the sheet on which the drawing is made, the type of paper, the margins, and other details relating to the making of the drawing. The reason for specifying the standards in detail is that the drawings are printed and published in a uniform style when the patent issues, and the drawings must also be such that they can be readily understood by persons using the patent descriptions.

Sets of Technical Drawings

Working Drawings: Working drawings are the set of technical drawings used during the manufacturing phase of a product. In architecture, these typically include civil drawings, architectural drawings, structural drawings, mechanical systems drawings, electrical drawings, and plumbing drawings.

Assembly Drawings

Assembly drawings show how different parts go together, identify those parts by number, and have a parts list, often referred to as a bill of materials.

Numerical Control

Numerical control (NC) refers to the automation of machine tools that are operated by abstractly programmed commands encoded on a storage medium, as opposed to manually controlled via handwheels or levers, or mechanically automated via cams alone. The first NC machines were built in the 1940s and 1950s, based on existing tools that were modified with motors that moved the controls to follow points fed into the system on punched tape. These early servomechanisms were rapidly augmented with analog and digital computers, creating the modern computer numerical control (CNC) machine tools that have revolutionised the machining processes.

In modern CNC systems, end-to-end component design is highly automated using computer-aided design (CAD) and computer-aided manufacturing (CAM) programmes. The programmes produce a computer file that is interpreted to extract the commands needed to operate a particular machine via a postprocessor, and then loaded into the CNC machines for production. Since any particular component might require the use of a number of different tools-drills, saws, etc., modern machines often combine multiple tools into a single "cell". In other cases, a number of different machines are used with an external controller and human or robotic operators that move the component from machine to machine. In either case, the complex series of steps needed to produce any part is highly automated and produces a part that closely matches the original CAD design.

Earlier Forms of Automation

Cams

The automation of machine tool control began in the 19th century with cams that "played" a machine tool in the way that cams had long been playing musical boxes or operating elaborate cuckoo clocks. Thomas Blanchard built his gun-stock-copying lathes (1820s–30s), and the work of people such as Christopher Miner Spencer developed the turret lathe into the screw machine (1870s). Cam-based automation had already reached a highly advanced state by World War I (1910s).

However, automation via cams is fundamentally different from numerical control because it cannot be abstractly programmed. Cams can encode information, but getting the information from the abstract level of an engineering drawing into the cam is a manual process that requires sculpting and/or machining and filing. Various forms of abstractly programmable control had existed during the 19th century: those of the Jacquard loom, player pianos, and mechanical computers pioneered by Charles Babbage and others. These developments had the potential for convergence with the automation of machine tool control starting in that century, but the convergence did not happen until many decades later.

Tracer Control

The application of hydraulics to cam-based automation resulted in tracing machines that used a stylus to trace a template, such as the enormous Pratt & Whitney "Keller Machine", which could copy templates several feet across. Another approach was "record and playback", pioneered at General Motors (GM) in the 1950s, which used a storage system to record the movements of a human machinist, and then play them back on demand. Analogous systems are common even today, notably the "teaching lathe" which gives new machinists a hands-on feel for the process. None of these were numerically programmable, however, and required a master machinist at some point in the process, because the "programming" was physical rather than numerical.

Servos and Selsyns

One barrier to complete automation was the required tolerances of the machining process, which are routinely on the order of thousandths of an inch. Although connecting some sort of control to a storage device like punched cards was easy, ensuring that the controls were moved to the correct position with the required accuracy was another issue. The movement of the tool resulted in varying forces on the controls that would mean a linear input would not result in linear tool motion. The key development in this area was the introduction of the servomechanism, which produced highly accurate measurement information. Attaching two servos together produced a selsyn, where a remote servo's motions were accurately matched by another. Using a variety of mechanical or electrical systems, the output of the selsyns could be read to ensure proper movement had occurred (in other words, forming a closed-loop control system).

The first serious suggestion that selsyns could be used for machining control was made by Ernst F. W. Alexanderson, a Swedish immigrant to the U.S. working at General Electric (GE). Alexanderson had worked on the problem of torque amplification that allowed the small output of a mechanical computer to drive very large motors, which GE used as part of a larger gun laying system for US Navy ships. Like machining, gun laying requires very high accuracies, much less than a degree, and the forces during the motion of the gun turrets was non-linear. In November 1931 Alexanderson suggested to the

Industrial Engineering Department that the same systems could be used to drive the inputs of machine tools, allowing it to follow the outline of a template without the strong physical contact needed by existing tools like the Keller Machine. He stated that it was a "matter of straight engineering development". However, the concept was ahead of its time from a business development perspective, and GE did not take the matter seriously until years later, when others had pioneered the field.

Parsons and the Invention of NC

The birth of NC is generally credited to John T. Parsons, a machinist and salesman at his father's machining company, Parsons Corp.

In 1942 he was told that helicopters were going to be the "next big thing" by the former head of Ford Trimotor production, Bill Stout. He called Sikorsky Aircraft to enquire about possible work, and soon got a contract to build the wooden stringers in the rotor blades. After setting up production at a disused furniture factory and ramping up production, one of the blades failed and it was traced to the spar. As at least some of the problem appeared to stem from spot welding a metal collar on the stringer to the metal spar, Parsons suggested a new method of attaching the stringers to the spar using adhesives, never before tried on an aircraft design.

That development led Parsons to consider the possibility of using stamped metal stringers instead of wood, which would be much stronger and easier to make. The stringers for the rotors were built from a design provided by Sikorsky, which was sent to Parsons as a series of 17 points defining the outline. Parsons then had to "fill in" the dots with a French curve to generate an outline they could use as a template to build the jigs for the wooden stringers. Making a metal cutting tool able to cut that particular shape proved to be difficult. Parsons went to Wright Field to see Frank Stulen, the head of the Propeller Lab Rotary Ring Branch. During their conversation, Stulen concluded that Parsons didn't really know what he was talking about. Parsons realised this, and hired Stulen on the spot. Stulen started work on 1 April 1946 and hired three new engineers to join him.

Stulen's brother worked at Curtis Wright Propeller, and mentioned that they were using punched card calculators for engineering calculations. Stulen decided to adopt the idea to run stress calculations on the rotors, the first detailed automated calculations on helicopter rotors. When Parsons saw what Stulen was doing with the punched card machines, he asked Stulen if they could be used to generate an outline with 200 points instead of the 17 they were given, and offset each point by the radius of a mill cutting tool. If you cut at each of those points, it would produce a relatively accurate cutout of the stringer even in hard steel, and it could easily be filed down to a smooth shape. The resulting tool would be useful as a template for stamping metal stringers. Stullen had no problem making such a programme, and used it to produce large tables of numbers that would be taken onto the machine floor. Here, one operator read the numbers off the charts to two other operators, one on each of the X- and Y-axes, and they would move the cutting head to that point and make a cut. This was called the "by-the-numbers method".

At that point Parsons conceived of a fully automated tool. With enough points on the outline, no manual working would be needed, but with manual operation, the time saved by having the part more closely match the outline was offset by the time needed to move the controls. If the machine's inputs were attached directly to the card reader, this delay, and any associated manual errors, would be removed and the number of points could be dramatically increased. Such a machine could repeatedly punch out perfectly accurate templates on command. But at the time Parsons had no funds to develop his ideas.

When one of Parsons's salesmen was on a visit to Wright Field, he was told of the problems the newly formed US Air Force was having with new jet designs. He asked if Parsons had anything to help them. Parsons showed Lockheed their idea of an automated mill, but they were uninterested. They decided to use 5-axis template copiers to produce the stringers, cutting from a metal template, and had already ordered the expensive cutting machine. But as Parsons noted:

Now just picture the situation for a minute. Lockheed had contracted to design a machine to make these wings. This machine had five axes of cutter movement, and each of these was tracer controlled using a template. Nobody was using my method of making templates, so just imagine what chance they were going to have of making an accurate airfoil shape with inaccurate templates.

Parson's worries soon came true, and Lockheed's protests that they could fix the problem eventually rang hollow. In 1949 the Air Force arranged funding for Parsons to build his machines on his own. Early work with Snyder Machine & Tool Corp proved the system of directly driving the controls from motors failed to give the accuracy needed to set the machine for a perfectly smooth cut. Since the mechanical controls did not respond in a linear fashion, you couldn't simply drive it with a given amount of power, because the differing forces meant the same amount of power would not always produce the same amount of motion in the controls. No matter how many points you included, the outline would still be rough.

Enter MIT

This was not an impossible problem to solve, but would require some sort of feedback system, like a selsyn, to directly measure how far the controls had actually turned. Faced with the daunting task of building such a system, in the spring of 1949 Parsons turned to Gordon S. Brown's Servomechanisms Laboratory at MIT, which was a world leader in mechanical computing and feedback systems. During the war the Lab had built a number of complex motor-driven devices like the motorised gun turret systems for the Boeing B-29 Superfortress and the automatic tracking system for the SCR-584 radar. They were naturally suited to technological transfer into a prototype of Parsons's automated "by-the-numbers" machine.

The MIT team was led by William Pease assisted by James McDonough. They quickly concluded that Parsons's design could be greatly improved; if the machine did not simply cut *at* points A and B, but instead moved smoothly *between* the points, then not only would it make a perfectly smooth cut, but could do so with many fewer points - the mill could cut lines directly instead of having to define a large number of cutting points to "simulate" it. A three-way agreement was arranged between Parsons, MIT, and the Air Force, and the project officially ran from July 1949 to June 1950. The contract called for the construction of two "Card-a-matic Milling Machines", a prototype and a production system. Both to be handed to Parsons for attachment to one of their mills in order to develop a deliverable system for cutting stringers.

Instead, in 1950 MIT bought a surplus Cincinnati Milling Machine Company "Hydro-Tel" mill of their own and arranged a new contract directly with the Air Force that froze Parsons out of further development. Parsons would later comment that he "never dreamed that anybody as reputable as MIT would deliberately go ahead and take over my project." In spite of the development being handed to MIT, Parsons filed for a patent on "Motor Controlled Apparatus for Positioning Machine Tool" on 5 May 1952, sparking a filing by MIT for a "Numerical Control Servo-System" on 14 August 1952. Parsons received US Patent 2,820,187 on 14 January 1958, and the company sold an exclusive license to Bendix. IBM, Fujitsu and General Electric all took sub-licenses after having already started development of their own devices.

MIT's Machine

MIT fit gears to the various handwheel inputs and drove them with roller chains connected to motors, one for each of the machine's three axes (X, Y, and Z). The associated controller consisted of five refrigerator-sized cabinets that, together, were almost as large as the mill they were connected to. Three of the cabinets contained the motor controllers, one controller for each motor, the other two the digital reading system.

Unlike Parsons's original punched card design, the MIT design used standard 7-track punch tape for input. Three of the tracks were used to control the different axes of the machine, while the other four encoded various control information. The tape was read in a cabinet that also housed six relay-based hardware registers, two for each axis. With every read operation the previously read point was copied into the "starting point" register, and the newly read one into the "ending point". The tape was read continually and the number in the register increased until a "stop" instruction was encountered, four holes in a line.

The final cabinet held a clock that sent pulses through the registers, compared them, and generated output pulses that interpolated between the points. For instance, if the points were far apart the output would have pulses with every clock cycle, whereas closely spaced points would only generate pulses after multiple clock cycles. The pulses are sent into a summing register in the motor controllers, counting up by the number of pulses every time they were received. The summing registers were connected to a digital to analog converter that increased power to the motors as the count in the registers increased.

The registers were decremented by encoders attached to the motors and the mill itself, which would reduce the count by one for every one degree of rotation. Once the second point was reached the pulses from the clock would stop, and the motors would eventually drive the mill to the encoded position. Each 1 degree rotation of the controls produced a 0.0005 inch movement of the cutting head. The programmer could control the speed of the cut by selecting points that were closer together for slow movements, or further apart for rapid ones.

The system was publicly demonstrated in September 1952, appearing in that month's *Scientific American*. MIT's system was an outstanding success by any technical measure, quickly making any complex cut with extremely high accuracy that could not easily be duplicated by hand. However, the system was terribly complex, including 250 vacuum tubes, 175 relays and numerous moving parts, reducing its reliability in a production environment. It was also very expensive, the total bill presented to the Air Force was $360,000.14, $2,641,727.63 in 2005 dollars. Between 1952 and 1956 the system was used to mill a number of one-off designs for various aviation firms, in order to study their potential economic impact.

Proliferation of NC

The Air Force funding for the project ran out in 1953, but development was picked up by the Giddings and Lewis Machine Tool Co. In 1955 many of the MIT team left to form Concord Controls, a commercial NC company with Giddings' backing, producing the Numericord controller. Numericord was similar to the MIT design, but replaced the punch tape with a magnetic tape reader that General Electric was working on. The tape contained a number of signals of different phases, which directly encoded the angle of the various controls. The tape was played at a constant speed in the controller, which set its half of the selsyn to the encoded angles while the remote side was attached to the machine controls. Designs were still encoded on paper tape, but the tapes were transferred to a reader/writer that converted them into magnetic form. The magtapes could then be used on any of the machines on the floor, where the controllers were greatly reduced in complexity. Developed to produce highly accurate dies for an aircraft skinning press, the Numericord "NC5" went into operation at G&L's plant at Fond du Lac, WI in 1955.

Monarch Machine Tool also developed an numerical controlled lathe, starting in 1952. They demonstrated their machine at the 1955 Chicago Machine Tool Show (predecessor of today's IMTS), along with a number of other vendors with punched card or paper tape machines that were either fully developed or in prototype form. These included Kearney & Trecker's Milwaukee-Matic II that could change its cutting tool under numerical control, a common feature on modern machines.

A Boeing report noted that "numerical control has proved it can reduce costs, reduce lead times, improve quality, reduce tooling and increase productivity." In spite of these developments, and glowing reviews from the few users, uptake of NC was relatively slow. As Parsons later noted:

The NC concept was so strange to manufacturers, and so slow to catch on, that the US Army itself finally had to build 120 NC machines and lease them to various manufacturers to begin popularising its use.

In 1958 MIT published its report on the economics of NC. They concluded that the tools were competitive with human operators, but simply moved the time from the machining to the creation of the tapes. In *Forces of Production*, Noble claims that this was the whole point as far as the Air Force was concerned; moving the process off of the highly unionised factory floor and into the un-unionised white collar design office. The cultural context of the early 1950s, a second Red Scare with a widespread fear of a bomber gap and of domestic subversion, sheds light on this interpretation. It was strongly feared that the West would lose the defence production race to the Communists, and that syndicalist power was a path towards losing, either by "getting too soft" (less output, greater unit expense) or even by Communist sympathy and subversion within unions (arising from their common theme of empowering the working class).

CNC Arrives

Many of the commands for the experimental parts were programmed "by hand" to produce the punch tapes that were used as input. During the development of Whirlwind, MIT's real-time computer, John Runyon coded a number of subroutines to produce these tapes under computer control. Users could enter a list of points and speeds, and the programme would generate the punch tape. In one instance, this process reduced the time required to produce the instruction list and mill the part from 8 hours to 15 minutes. This led to a proposal to the Air Force to produce a generalised "programming" language for numerical control, which was accepted in June 1956.

Starting in September, Ross and Pople outlined a language for machine control that was based on points and lines, developing this over several years into the APT programming language. In 1957 the Aircraft Industries Association (AIA) and Air Material Command at Wright-Patterson Air Force Base joined with MIT to standardise this work and produce a fully computer-controlled NC system. On 25 February 1959 the combined team held a press conference showing the results, including a 3D machined aluminium ash tray that was handed out in the press kit.

Meanwhile, Patrick Hanratty was making similar developments at GE as part of their partnership with G&L on the Numericord. His language, PRONTO, beat APT into commercial use when it was released in 1958. Hanratty then went on to develop MICR magnetic ink characters that were used in cheque processing, before moving to General Motors to work on the groundbreaking DAC-1 CAD system.

APT was soon extended to include "real" curves in 2D-APT-II. With its release, MIT reduced its focus on CNC as it moved into CAD experiments. APT development was picked up with the AIA in San Diego, and in 1962, by Illinois Institute of Technology Research. Work on making APT an international standard started in 1963 under USASI X3.4.7, but many manufacturers of CNC machines had their own one-off additions (like PRONTO), so standardisation was not completed until 1968, when there were 25 optional add-ins to the basic system.

Just as APT was being released in the early 1960s, a second generation of lower-cost transistorised computers was hitting the market that were able to process much larger volumes of information in production settings. This reduced the cost of implementing a NC system and by the mid 1960s, APT runs accounted for a third of all computer time at large aviation firms.

CAD Meets CNC

While the Servomechanisms Lab was in the process of developing their first mill, in 1953, MIT's Mechanical Engineering Department dropped the requirement that undergraduates take courses in drawing. The instructors formerly teaching these programmes were merged into the Design Division, where an informal discussion of computerised design started. Meanwhile the Electronic Systems Laboratory, the newly rechristened Servomechanisms Laboratory, had been discussing whether or not design would ever start with paper diagrams in the future.

In January 1959, an informal meeting was held involving individuals from both the Electronic Systems Laboratory and the Mechanical Engineering Department's Design Division. Formal meetings followed in April and May, which resulted in the "Computer-Aided Design Project". In December 1959, the Air Force issued a one year contract to ESL for $223,000 to fund the project, including $20,800 earmarked for 104 hours of computer time at $200 per hour. This proved to be far too little for the ambitious programme they had in mind, although their engineering calculation system, AED, was released in March 1965.

In 1959, General Motors started an experimental project to digitise, store and print the many design sketches being generated in the various GM design departments. When the basic concept demonstrated that it could work, they started the DAC-1 project with IBM to develop a production version. One part of the DAC project was the direct conversion of paper diagrams into 3D models, which were then converted into APT commands and cut on milling machines. In November 1963 a trunk lid design moved from 2D paper sketch to 3D clay prototype for the first time. With the exception of the initial sketch, the design-to-production loop had been closed.

Meanwhile, MIT's offsite Lincoln Labs was building computers to test new transistorised designs. The ultimate goal was essentially a transistorised Whirlwind known as TX-2, but in order to test various circuit designs a smaller version known as TX-0 was built first. When construction of TX-2 started, time in TX-0 freed up and this led to a number of experiments involving interactive input and use of the machine's CRT display for graphics. Further development of these concepts led to Ivan Sutherland's groundbreaking Sketchpad programme on the TX-2.

Sutherland moved to the University of Utah after his Sketchpad work, but it inspired other MIT graduates to attempt the first true CAD system. It was Electronic Drafting Machine (EDM), sold to Control Data and known as "Digigraphics", that Lockheed used to build production parts for the C-5 Galaxy, the first example of an end-to-end CAD/CNC production system.

By 1970 there were a wide variety of CAD firms including Intergraph, Applicon, Computervision, Auto-trol Technology, UGS Corp. and others, as well as large vendors like CDC and IBM.

Proliferation of CNC

The price of computer cycles fell drastically during the 1960s with the widespread introduction of useful minicomputers. Eventually it became less expensive to handle the motor control and feedback with a computer programme than it was with dedicated servo systems. Small computers were dedicated to a single mill, placing the entire process in a small box. PDP-8's and Data General Nova computers were common in these roles. The introduction of the microprocessor in the 1970s further reduced the cost of implementation, and today almost all CNC machines use some form of microprocessor to handle all operations.

The introduction of lower-cost CNC machines radically changed the manufacturing industry. Curves are as easy to cut as straight lines, complex 3-D structures are relatively easy to produce, and the number of machining steps that required human action have been dramatically reduced. With the increased automation of manufacturing processes with CNC machining, considerable improvements in consistency and quality have been achieved with no strain on the operator. CNC automation reduced the frequency of errors and provided CNC operators with time to perform additional tasks. CNC automation also allows for more flexibility in the way parts are held in the manufacturing process and the time required to change the machine to produce different components.

During the early 1970s the Western economies were mired in slow economic growth and rising employment costs, and NC machines started to become more attractive. The major U.S. vendors were slow to respond to the demand for machines suitable for lower-cost NC systems, and into this void stepped the Germans. In 1979, sales of German machines surpassed the U.S. designs for the first time. This cycle quickly repeated itself, and by 1980 Japan had taken a leadership position, U.S. sales dropping all the time. Once sitting in the #1 position in terms of sales on a top-ten chart consisting entirely of U.S. companies in 1971, by 1987 Cincinnati Milacron was in 8th place on a chart heavily dominated by Japanese firms.

Many researchers have commented that the U.S. focus on high-end applications left them in an uncompetitive situation when the economic downturn in the early 1970s led to greatly increased demand for low-cost NC systems. Unlike the U.S. companies, who had focused on the highly profitable aerospace market, German and Japanese manufacturers targeted lower-profit segments from the start and were able to enter the low-cost markets much more easily.

As computing and networking evolved, so did direct numerical control (DNC). Its long-term coexistence with less networked variants of NC and CNC is explained by the fact that individual firms tend to stick with whatever is profitable, and their time and money for trying out alternatives is limited. This explains why machine tool models and tape storage media persist in grandfathered fashion even as the state of the art advances.

DIY, Hobby, and Personal CNC

Recent developments in small scale CNC have been enabled, in large part, by the Enhanced Machine Controller project from the National Institute of Standards and Technology (NIST), an agency of the US Government's Department of Commerce. EMC is a public domain programme operating under the Linux operating system and working on PC based hardware.

After the NIST project ended, development continued, leading to EMC2 which is licensed under the GNU General Public License and Lesser GNU General Public License (GPL and LGPL). Derivations of the original EMC software have also led to several proprietary PC based programmes notably TurboCNC, and Mach3, as well as embedded systems based on proprietary hardware. The availability of these PC based control programmes has led to the development of DIY CNC, allowing hobbyists to build their own using open source hardware designs. The same basic architecture has allowed manufacturers, such as Sherline and Taig, to produce turnkey lightweight desktop milling machines for hobbyists.

The easy availability of PC based software and support information of Mach3, written by Art Fenerty, lets anyone with some time and technical expertise make complex parts for home and prototype use. Fenerty is considered a principal founder of Windows-based PC CNC machining.

Eventually, the homebrew architecture was fully commercialised and used to create larger machinery suitable for commercial and industrial applications. This class of equipment has been referred to as Personal CNC. Parallel to the evolution of personal computers, Personal CNC has its roots in EMC and PC based control, but has evolved to the point where it can replace larger conventional equipment in many instances. As with the Personal Computer, Personal CNC is characterised by equipment whose size, capabilities, and original sales price make it useful for individuals, and which is intended to be operated directly by an end user, often without professional training in CNC technology.

Today

Although modern data storage techniques have moved on from punch tape in almost every other role, tapes are still relatively common in CNC systems. Several reasons explain this. One is easy backward compatibility of existing programmes. Companies were spared the trouble of re-writing existing tapes into a new format. Another is the principle, mentioned earlier, that individual firms tend to stick with whatever is profitable, and their time and money for trying out alternatives is limited. A small firm that has found a profitable niche may keep older equipment in service for years because “if it ain't broke, don't fix it.” Competition places natural limits on that approach, as some amount of innovation and continuous improvement eventually becomes necessary, lest competitors be the ones who find the way to the "better mousetrap".

One change that was implemented fairly widely was the switch from paper to mylar tapes, which are much more mechanically robust. Floppy disks, USB flash drives and local area networking have replaced the tapes to some degree, especially in larger environments that are highly integrated.

The proliferation of CNC led to the need for new CNC standards that were not encumbered by licensing or particular design concepts, like APT. A number of different "standards" proliferated for a time, often based around vector graphics markup languages supported by plotters. One such standard has since become very common, the "G-code" that was originally used on Gerber Scientific plotters and then adapted for CNC use. The file format became so widely used that it has been embodied in an EIA standard. In turn, while G-code is the predominant language used by CNC machines today, there is a push to supplant it with STEP-NC, a system that was deliberately designed for CNC, rather than grown from an existing plotter standard.

While G-code is the most common method of programming, some machine-tool/control manufacturers also have invented their own proprietary "conversational" methods of programming, trying to make it easier to programme simple parts and make set-up and modifications at the machine easier (such as Mazak's Mazatrol and Hurco). These have met with varying success.

A more recent advancement in CNC interpreters is support of logical commands, known as parametric programming (also known as macro programming). Parametric programmes include both device commands as well as a control language similar to BASIC. The programmer can make if/then/else statements, loops, subprogram calls, perform various arithmetic, and manipulate variables to create a large degree of freedom within one programme. An entire product line of different sizes can be programmed using logic and simple math to create and scale an entire range of parts, or create a stock part that can be scaled to any size a customer demands.

Since about 2006, the idea has been suggested and pursued to foster the convergence with CNC and DNC of several trends elsewhere in the world of information technology that have not yet much affected CNC and DNC. One of these trends is the combination of greater data collection (more sensors), greater and more automated data exchange (via building new, open industry-standard XML schemas), and data mining to yield a new level of business intelligence and workflow automation in manufacturing. Another of these trends is the emergence of widely published APIs together with the aforementioned open data standards to encourage an ecosystem of user-generated apps and mashups, which can be both open and commercial—in other words, taking the new IT culture of app marketplaces that began in web development and smartphone app development and spreading it to CNC, DNC, and the other factory automation systems that are networked with the CNC/DNC. MTConnect is a leading effort to bring these ideas into successful implementation.

Description

Modern CNC mills differ little in concept from the original model built at MIT in 1952. Mills typically consist of a table that moves in the X and Y axes, and a tool spindle that moves in the Z (depth). The position of the tool is driven by motors through a series of step-down gears in order to provide highly accurate movements, or in modern designs, direct-drive stepper motors. Closed-loop control is not mandatory today, as open-loop control works as long as the forces are kept small enough.

As the controller hardware evolved, the mills themselves also evolved. One change has been to enclose the entire mechanism in a large box as a safety measure, often with additional safety interlocks to ensure the operator is far enough from the working piece for safe operation. Most new CNC systems built today are completely electronically controlled. CNC-like systems are now used for any process that can be described as a series of movements and operations. These include laser cutting, welding, friction stir welding, ultrasonic welding, flame and plasma cutting, bending, spinning, pinning, gluing, fabric cutting, sewing, tape and fibre placement, routing, picking and placing (PnP), and sawing.

Tools with CNC Variants

- Drills
- EDMs
- Lathes
- Milling machines
- Wood routers
- Sheet metal works (Turret Punch)
- Wire bending machines

- Hot-wire foam cutters
- Plasma cuttings
- Water jet cutters
- Laser cutting
- Oxy-fuel
- Surface grinders
- Cylindrical grinders
- 3D Printing
- Induction hardening machines.

Tool / Machine Crashing

In CNC, a "crash" occurs when the machine moves in such a way that is harmful to the machine, tools, or parts being machined, sometimes resulting in bending or breakage of cutting tools, accessory clamps, vises, and fixtures, or causing damage to the machine itself by bending guide rails, breaking drive screws, or causing structural components to crack or deform under strain. A mild crash may not damage the machine or tools, but may damage the part being machined so that it must be scrapped.

Many CNC tools have no inherent sense of the absolute position of the table or tools when turned on. They must be manually "homed" or "zeroed" to have any reference to work from, and these limits are just for figuring out the location of the part to work with it, and aren't really any sort of hard motion limit on the mechanism. It is often possible to drive the machine outside the physical bounds of its drive mechanism, resulting in a collision with itself or damage to the drive mechanism.

Many CNC tools also don't know anything about their working environment. They often lack any form of sensory capability to detect problems with the machining process, and will not abort if something goes wrong. They blindly follow the machining code provided and it is up to an operator to detect if a crash is either occurring or about to occur, and for the operator to manually abort the cutting process.

If the drive system is weaker than the machine structural integrity, then the drive system simply pushes against the obstruction and the drive motors "slip in place". The machine tool may not detect the collision or the slipping, so for example the tool should now be at 210mm on the X axis but is in fact at 32mm where it hit the obstruction and kept slipping. All of the next tool motions will be off by -178mm on the X axis, and all future motions are now invalid, which may result in further collisions with clamps, vises, or the machine itself.

Collision detection and avoidance is possible, through the use of absolute position sensors (optical encoder strips or disks) to verify that motion occurred, or torque sensors or power-draw sensors on the drive system to detect abnormal strain when the machine should just be moving and not cutting, but these are not a common component of most CNC tools.

Instead, most CNC tools simply rely on the assumed accuracy of stepper motors that rotate a specific number of degrees in response to magnetic field changes. It is often assumed the stepper is perfectly accurate and never mis-steps, so tool position monitoring simply involves counting the number of pulses sent to the stepper over time. An alternate means of stepper position monitoring is usually not available, so crash or slip detection is not possible.

Numerical Accuracy US Equipment Backlash

Within the numerical systems of CNC programming it is possible for the code generator to assume that the controlled mechanism is always perfectly accurate, or that accuracy tolerances are identical for all cutting or movement directions. This is not always a true condition of CNC tools.

CNC tools with a large amount of mechanical backlash can still be highly accurate if the drive or cutting mechanism is only driven so as to apply cutting force from one direction, and all driving systems are pressed tight together in that one cutting direction. However a CNC device with high backlash and a dull cutting tool can lead to cutter chatter and possible workpiece gouging. Backlash also affects accuracy of some operations involving axis movement reversals during cutting, such as the milling of a circle, where axis motion is sinusoidal. However, this can be compensated for if the amount of backlash is precisely known by linear encoders or manual measurement.

The high backlash mechanism itself is not necessarily relied on to be repeatably accurate for the cutting process, but some other reference object or precision surface may be used to zero the mechanism, by tightly applying pressure against the reference and setting that as the zero reference for all following CNC-encoded motions. This is similar to the manual machine tool method of clamping a micrometer onto a reference beam and adjusting the Vernier dial to zero using that object as the reference.

Frontiers of Research

Mechanical engineers are constantly pushing the boundaries of what is physically possible in order to produce safer, cheaper, and more efficient machines and mechanical systems. Some technologies at the cutting edge of mechanical engineering are listed below.

Micro Electro-Mechanical Systems (MEMS)

Micron-scale mechanical components such as springs, gears, fluidic and heat transfer devices are fabricated from a variety of substrate materials such as silicon, glass and polymers like SU8. Examples of MEMS components are the accelerometers that are used as car airbag sensors, modern cell phones, gyroscopes for precise positioning and microfluidic devices used in biomedical applications.

Friction Stir Welding

Friction-stir welding (FSW) is a solid-state joining process (meaning the metal is not melted during the process) and is used for applications where the original metal characteristics must remain unchanged as far as possible. This process is primarily used on aluminium, and most often on large pieces which cannot be easily heat treated post weld to recover temper characteristics.

It was invented and experimentally proven by Wayne Thomas and a team of his colleagues at The Welding Institute UK in December 1991. TWI holds a number of patents on the process, the first being the most descriptive.

Principle of Operation

In FSW, a cylindrical-shouldered tool, with a profiled threaded/ unthreaded probe (nib or pin) is rotated at a constant speed and fed at a constant traverse rate into the joint line between two pieces of sheet or plate material, which are butted together. The parts have to be clamped rigidly onto a backing bar in a manner that prevents the abutting joint faces from being forced apart. The length of the nib is slightly less than the weld depth required and the tool shoulder should be in intimate contact with the work surface. The nib is then moved against the work, or vice versa.

Frictional heat is generated between the wear-resistant welding tool shoulder and nib, and the material of the work pieces. This heat, along with the heat generated by the mechanical mixing process and the adiabatic heat within the material, cause the stirred materials to soften without reaching the melting point (hence cited as a solid-state process), allowing the traversing of the tool along the weld line in a plasticised tubular shaft of metal. As the pin is moved in the direction of welding, the leading face of the pin, assisted by a special pin profile, forces plasticised material to the back of the pin while applying a substantial forging force to consolidate the weld metal. The welding of the material is facilitated by severe plastic deformation in the solid state, involving dynamic recrystallisation of the base material.

Microstructural Features

The solid-state nature of the FSW process, combined with its unusual tool and asymmetric nature, results in a highly characteristic microstructure. The microstructure can be broken up into the following zones:

- The stir zone (also nugget, dynamically recrystallised zone) is a region of heavily deformed material that roughly corresponds to the location of the pin during welding. The grains within the stir zone are roughly equiaxed and often an order of magnitude smaller than the grains in the parent material. A unique feature of the stir zone is the common occurrence of several concentric rings which has been referred to as an "onion-ring" structure. The precise origin of these rings has not been firmly established, although variations in particle number density, grain size and texture have all been suggested.
- The flow arm zone is on the upper surface of the weld and consists of material that is dragged by the shoulder from the retreating side of the weld, around the rear of the tool, and deposited on the advancing side.
- The thermo-mechanically affected zone (TMAZ) occurs on either side of the stir zone. In this region the strain and temperature are lower and the effect of welding on the microstructure is correspondingly smaller. Unlike the stir zone the microstructure is recognizably that of the parent material, albeit significantly deformed and rotated. Although the term TMAZ technically refers to the entire deformed region it is often used to describe any region not already covered by the terms stir zone and flow arm.
- The heat-affected zone (HAZ) is common to all welding processes. As indicated by the name, this region is subjected to a thermal cycle but is not deformed during welding. The temperatures are lower than those in the TMAZ but may still have a significant effect if the microstructure is thermally

unstable. In fact, in age-hardened aluminium alloys this region commonly exhibits the poorest mechanical properties.

Advantages and Limitations

The solid-state nature of FSW immediately leads to several advantages over fusion welding methods since any problems associated with cooling from the liquid phase are immediately avoided. Issues such as porosity, solute redistribution, solidification cracking and liquation cracking are not an issue during FSW. In general, FSW has been found to produce a low concentration of defects and is very tolerant to variations in parameters and materials.

Nevertheless, FSW is associated with a number of unique defects. Insufficient weld temperatures, due to low rotational speeds or high traverse speeds, for example, mean that the weld material is unable to accommodate the extensive deformation during welding. This may result in long, tunnel-like defects running along the weld which may occur on the surface or subsurface. Low temperatures may also limit the forging action of the tool and so reduce the continuity of the bond between the material from each side of the weld. The light contact between the material has given rise to the name "kissing-bond". This defect is particularly worrying since it is very difficult to detect using non-destructive methods such as X-ray or ultrasonic testing. If the pin is not long enough or the tool rises out of the plate then the interface at the bottom of the weld may not be disrupted and forged by the tool, resulting in a lack-of-penetration defect. This is essentially a notch in the material which can be a potent source of fatigue cracks.

A number of potential advantages of FSW over conventional fusion-welding processes have been identified:

- Good mechanical properties in the as welded condition
- Improved safety due to the absence of toxic fumes or the spatter of molten material.
- No consumables — A threaded pin made of conventional tool steel, e.g., hardened H13, can weld over 1000m of aluminium, and no filler or gas shield is required for aluminium.
- Easily automated on simple milling machines — lower setup costs and less training.
- Can operate in all positions (horizontal, vertical, etc.), as there is no weld pool.

- Generally good weld appearance and minimal thickness under/over-matching, thus reducing the need for expensive machining after welding.
- Low environmental impact.

However, some disadvantages of the process have been identified:

- Exit hole left when tool is withdrawn.
- Large down forces required with heavy-duty clamping necessary to hold the plates together.
- Less flexible than manual and arc processes (difficulties with thickness variations and non-linear welds).
- Often slower traverse rate than some fusion welding techniques, although this may be offset if fewer welding passes are required.

Important Welding Parameters

Tool Rotation and Traverse Speeds: There are two tool speeds to be considered in friction-stir welding; how fast the tool rotates and how quickly it traverses the interface. These two parameters have considerable importance and must be chosen with care to ensure a successful and efficient welding cycle. The relationship between the welding speeds and the heat input during welding is complex but, in general, it can be said that increasing the rotation speed or decreasing the traverse speed will result in a hotter weld. In order to produce a successful weld it is necessary that the material surrounding the tool is hot enough to enable the extensive plastic flow required and minimise the forces acting on the tool. If the material is too cold then voids or other flaws may be present in the stir zone and in extreme cases the tool may break.

Excessively high heat input, on the other hand may be detrimental to the final properties of the weld. Theoretically, this could even result in defects due to the liquation of low-melting-point phases (similar to liquation cracking in fusion welds). These competing demands lead onto the concept of a "processing window": the range of processing parameters viz. tool rotation and traverse speed, that will produce a good quality weld. Within this window the resulting weld will have a sufficiently high heat input to ensure adequate material plasticity but not so high that the weld properties are excessively deteriorated.

Tool Tilt and Plunge Depth

The plunge depth is defined as the depth of the lowest point of the shoulder below the surface of the welded plate and has been found to be a critical parameter for ensuring weld quality. Plunging the shoulder below the plate surface increases the pressure below the tool and helps ensure adequate forging of the material at the rear of the tool. Tilting the tool by 2-4 degrees, such that the rear of the tool is lower than the front, has been found to assist this forging process. The plunge depth needs to be correctly set, both to ensure the necessary downward pressure is achieved and to ensure that the tool fully penetrates the weld. Given the high loads required the welding machine may deflect and so reduce the plunge depth compared to the nominal setting, which may result in flaws in the weld. On the other hand an excessive plunge depth may result in the pin rubbing on the backing plate surface or a significant undermatch of the weld thickness compared to the base material. Variable load welders have been developed to automatically compensate for changes in the tool displacement while TWI have demonstrated a roller system that maintains the tool position above the weld plate.

Tool Design

The design of the tool is a critical factor as a good tool can improve both the quality of the weld and the maximum possible welding speed.

It is desirable that the tool material is sufficiently strong, tough and hard wearing, at the welding temperature. Further it should have a good oxidation resistance and a low thermal conductivity to minimise heat loss and thermal damage to the machinery further up the drive train. Hot-worked tool steel such as AISI H13 has proven perfectly acceptable for welding aluminium alloys within thickness ranges of 0.5 – 50 mm but more advanced tool materials are necessary for more demanding applications such as highly abrasive metal matrix composites or higher melting point materials such as steel or titanium.

Improvements in tool design have been shown to cause substantial improvements in productivity and quality. TWI has developed tools specifically designed to increase the depth of penetration and so increase the plate thickness that can be successfully welded. An example is the "whorl" design that uses a tapered pin with re-entrant features or a variable pitch thread in order to improve the downwards flow of material. Additional designs include the Triflute and Trivex series. The Triflute design has a complex system of three tapering, threaded re-entrant flutes that appear to increase material movement around the tool. The Trivex tools use a simpler, non-cylindrical, pin and have been found to reduce the forces acting on the tool during welding.

The majority of tools have a concave shoulder profile which acts as an escape volume for the material displaced by the pin, prevents material from extruding out of the sides of the shoulder and maintains downwards pressure and hence good forging of the material behind the tool. The Triflute tool uses an alternative system with a series of concentric grooves machined into the surface which are intended to produce additional movement of material in the upper layers of the weld. Widespread commercial applications of friction stir welding process for steels and other hard alloys such as titanium alloys will require the development of cost effective and durable tools. Material selection, design and cost are important considerations in the search for commercially useful tools for the welding of hard materials. Work is continuing to better understand the effects of tool material's composition. Structure, properties and geometry on their performance, durability and cost.

Welding Forces

During welding a number of forces will act on the tool:

- A downwards force is necessary to maintain the position of the tool at or below the material surface. Some friction-stir welding machines operate under load control but in many cases the vertical position of the tool is preset and so the load will vary during welding.
- The traverse force acts parallel to the tool motion and is positive in the traverse direction. Since this force arises as a result of the resistance of the material to the motion of the tool it might be expected that this force will decrease as the temperature of the material around the tool is increased.
- The lateral force may act perpendicular to the tool traverse direction and is defined here as positive towards the advancing side of the weld.
- Torque is required to rotate the tool, the amount of which will depend on the down force and friction coefficient (sliding friction) and/or the flow strength of the material in the surrounding region (sticking friction).

In order to prevent tool fracture and to minimise excessive wear and tear on the tool and associated machinery, the welding cycle should be modified so that the forces acting on the tool are as low as possible, and abrupt changes are avoided. In order to find the best combination of welding parameters it is likely that a compromise must be reached, since the conditions that favour low forces (e.g. high heat input, low travel speeds) may be undesirable from the point of view of productivity and weld properties.

Flow of Material

Early work on the mode of material flow around the tool used inserts of a different alloy, which had a different contrast to the normal material when viewed through a microscope, in an effort to determine where material was moved as the tool passed. The data was interpreted as representing a form of in-situ extrusion where the tool, backing plate and cold base material form the "extrusion chamber" through which the hot, plasticised material is forced. In this model the rotation of the tool draws little or no material around the front of the pin instead the material parts in front of the pin and passes down either side. After the material has passed the pin the side pressure exerted by the "die" forces the material back together and consolidation of the join occurs as the rear of the tool shoulder passes overhead and the large down force forges the material.

More recently, an alternative theory has been advanced that advocates considerable material movement in certain locations. This theory holds that some material does rotate around the pin, for at least one rotation, and it is this material movement that produces the "onion-ring" structure in the stir zone. The researchers used a combination of thin Cu strip inserts and a "frozen pin" technique, where the tool is rapidly stopped in place. They suggested that material motion occurs by two processes:

1. Material on the advancing front side of a weld enters into a zone that rotates and advances with the pin. This material was very highly deformed and sloughs off behind the pin to form arc-shaped features when viewed from above (i.e. down the tool axis). It was noted that the copper entered the rotational zone around the pin, where it was broken up into fragments. These fragments were only found in the arc shaped features of material behind the tool.
2. The lighter material came from the retreating front side of the pin and was dragged around to the rear of the tool and filled in the gaps between the arcs of advancing side material. This material did not rotate around the pin and the lower level of deformation resulted in a larger grain size.

The primary advantage of this explanation is that it provides a plausible explanation for the production of the onion-ring structure.

The marker technique for friction stir welding provides data on the initial and final positions of the marker in the welded material. The flow of material is then reconstructed from these positions. Detailed material flow field during friction stir welding can also be calculated from theoretical considerations based on fundamental scientific principles. Material flow calculations are routinely used in numerous engineering applications. Calculation of material flow fields in friction stir welding can be undertaken both using comprehensive numerical simulations or simple but insightful analytical equations. The comprehensive models for the calculation of material flow fields also provide important information such as geometry of the stir zone and the torque on the tool. The numerical simulations have shown the ability to correctly predict the results from marker experiments and the stir zone geometry observed in friction stir welding experiments.

Generation and Flow of Heat

For any welding process it is, in general, desirable to increase the travel speed and minimise the heat input as this will increase productivity and possibly reduce the impact of welding on the mechanical properties of the weld. At the same time it is necessary to ensure that the temperature around the tool is sufficiently high to permit adequate material flow and prevent flaws or tool fracture.

When the traverse speed is increased, for a given heat input, there is less time for heat to conduct ahead of the tool and the thermal gradients are larger. At some point the speed will be so high that the material ahead of the tool will be too cold, and the flow stress too high, to permit adequate material movement, resulting in flaws or tool fracture. If the "hot zone" is too large then there is scope to increase the traverse speed and hence productivity.

The welding cycle can be split into several stages during which the heat flow and thermal profile will be different:

- *Dwell:* The material is preheated by a stationary, rotating tool in order to achieve a sufficient temperature ahead of the tool to allow the traverse. This period may also include the plunge of the tool into the workpiece.
- *Transient Heating:* When the tool begins to move there will be a transient period where the heat production and temperature around the tool will alter in a complex manner until an essentially steady-state is reached.

- *Pseudo Steady-state:* Although fluctuations in heat generation will occur the thermal field around the tool remains effectively constant, at least on the macroscopic scale.
- *Post Steady-state:* Near the end of the weld heat may "reflect" from the end of the plate leading to additional heating around the tool.

Heat generation during friction-stir welding arises from two main sources: friction at the surface of the tool and the deformation of the material around the tool. The heat generation is often assumed to occur predominantly under the shoulder, due to its greater surface area, and to be equal to the power required to overcome the contact forces between the tool and the workpiece. The contact condition under the shoulder can be described by sliding friction, using a friction coefficient μ and interfacial pressure P, or sticking friction, based on the interfacial shear strength &tor; at an appropriate temperature and strain rate. Mathematical approximations for the total heat generated by the tool shoulder Q_{total} have been developed using both sliding and sticking friction models :

$$\text{(Sliding)}\ Q_{total} = \frac{2}{3}\pi P\mu\omega\left(R^3_{shoulder} - R^3_{pin}\right)$$

$$\text{(Sticking)}\ Q_{total} = \frac{2}{3}\pi\tau\omega\left(R^3_{shoulder} - R^3_{pin}\right)$$

where ω is the angular velocity of the tool, $R_{shoulder}$ is the radius of the tool shoulder and R_{pin} that of the pin. Several other equations have been proposed to account for factors such as the pin but the general approach remains the same.

A major difficulty in applying these equations is determining suitable values for the friction coefficient or the interfacial shear stress. The conditions under the tool are both extreme and very difficult to measure. To date, these parameters have been used as "fitting parameters" where the model works back from measured thermal data to obtain a reasonable simulated thermal field. While this approach is useful for creating process models to predict, for example, residual stresses it is less useful for providing insights into the process itself.

Applications

The FSW process is currently patented by TWI in most industrialised countries and licensed for over 183 users. Friction stir welding and its variants friction stir spot welding and friction stir processing are used for the following industrial applications:

Shipbuilding and Offshore

Two Scandinavian aluminium extrusion companies were in 1996 the first, who applied FSW commercially to the manufacture of fish freezer panels at Sapa, as well as deck panels and helicopter landing platforms at Marine Aluminium Aanensen, which subsequently merged with Hydro Aluminium Maritime to become Hydro Marine Aluminium. Some of these freezer panels are now also produced by Riftec and Bayards. In 1997 two-dimensional friction stir welds in the hydrodynamically flared bow section of the hull of the ocean viewer vessel The Boss were produced at Research Foundation Institute with the first portable FSW machine. The Super Liner Ogasawara at Mitsui Engineering and Shipbuilding is the largest friction stir welded ship so far. The Sea Fighter of Nichols Bros and the Freedom class Littoral Combat Ships contain prefabricated panels by the FSW fabricators Advanced Technology and Friction Stir Link, Inc. respectively. The Houbei class missile boat has friction stir welded rocket launch containers of China Friction Stir Centre. The HMNZS Rotoiti in New Zealand has FSW panels made by Donovans in a converted milling machine. Various companies apply FSW to armor plating for amphibious assault ships.

Aerospace

Boeing applies FSW to the Delta II and Delta IV expendable launch vehicles, and the first of these with a friction stir welded Interstage module has been launched in 1999. The process is also used for the Space Shuttle external tank, for Ares I and for the Orion Crew Vehicle test at NASA as well as Falcon 1 and Falcon 9 rockets at SpaceX. The toe nails for ramp of Boeing C-17 Globemaster III cargo aircraft by Advanced Joining Technologies and the cargo barrier beams for the Boeing 747 Large Cargo Freighter were the first commercially produced aircraft parts. FAA approved wings and fuselage panels of the Eclipse 500 aircraft were made at Eclipse Aviation, and this company delivered 259 friction stir welded business jets, before they were forced. Floor panels for Airbus A400M military aircraft are now made by Pfalz Flugzeugwerke and Embraer used FSW for the Legacy 450 and 500 Jets.

Automotive

Aluminium engine cradles and suspension struts for stretched Lincoln Town Car were the first automotive parts that were friction stir at Tower Automotive, who use the process also for the engine tunnel of the Ford GT. A spin-off of this company is called Friction Stir Link, Inc. and successfully exploits the FSW process, e.g. for the flatbed trailer "Revolution" of Fontaine Trailers. In Japan FSW is applied to suspension struts at Showa Denko and for joining of aluminium sheets to galvanised steel brackets for the boot lid of the Mazda MX-5. Friction stir spot welding is successfully used for the bonnet and rear doors of the Mazda RX-8 and the boot lid of the Toyota Prius. Wheels are friction stir welded at Simmons Wheels, UT Alloy Works and Fundo Rear seats for the Volvo V70 are friction stir welded at Sapa, HVAC pistons at Halla Climate Control and exhaust gas recirculation coolers at Pierburg. Tailor welded blanks are friction stir welded for the Audi R8 at Riftec. The B-column of the Audi R8 Spider is friction stir welded from two extrusions at Hammerer Aluminium Industries in Austria.

Railway Rolling Stock

Since 1997 roof panels were made from aluminium extrusions at Hydro Marine Aluminium with a bespoke 25m long FSW machine, e.g. for DSB class SA-SD trains of Alstom LHB Curved side and roof panels for the Victoria Line trains of London Underground, side panels for Bombardier's Electrostar trains at Sapa Group and side panels for Alstom's British Rail Class 390 Pendolino trains are made at Sapa Group Japanese commuter and express A-trains and British Rail Class 395 trains are friction stir welded by Hitachi, while Kawasaki applies friction stir spot welding to roof panels and Sumitomo Light Metal produces Shinkansen floor panels. Innovative FSW floor panels are made by Hammerer Aluminium Industries in Austria for the Stadler DOSTO double decker rail cars, to obtain an internal height of 2 m on both floors.

Heat sinks for cooling high-power electronics of locomotives are made at Sykatek, EBG, Austerlitz Electronics, EuroComposite, Sapa and Rapid Technic, and are the most common application of FSW due to the excellent heat transfer. The FSW process is also used for IGBT coolers at Sapa Group.

The lids of 50mm thick copper canisters for nuclear waste are attached to the cylinder by friction stir welding at SKB

Fabrication

Façade panels and athode sheets are friction stir welded at AMAG and Hammerer Aluminium Industries including friction stir lap welds of copper to aluminium. Bizerba#s meat slicers, Ökolüfter HVAC units and Siemens X-ray vacuum vessels are friction stir welded at Riftec. Vacuum valves and vessels are made by FSW at Japanese and Swiss companies. FSW is also used for the encapsulation of nuclear waste at SKB in 50mm thick copper canisters. Pressure vessels from ø1m semispherical forgings of 38.1mm thick aluminium alloy 2219 at Advanced Joining Technologies and Lawrence Livermore Nat Lab. Friction stir processing is applied to ship propellers at Friction Stir Link, Inc. and to hunting knives by DiamondBlade.

Composite Material

Additional citations for verification. Please help improve by adding citations to reliable sources. Unsourced material may be challenged and removed. (November 2008)

It has been suggested that Spread tow fabric and Technical fabric be merged into section. (Discuss) Proposed since June 2010.

A cloth of woven carbon fibre filaments, a common element in composite materials.

Composite materials, often shortened to composites or called composition materials, are engineered or naturally occurring materials made from two or more constituent materials with significantly different physical or chemical properties which remain separate and distinct at the macroscopic or microscopic scale within the finished structure.

A common example of a composite would be disc brake pads, which consists of hard ceramic particles embedded in soft metal matrix. Another example is found in shower stalls and bathtubs which are made of fibreglass. Imitation granite and cultured marble sinks and countertops are also widely used. The most advanced examples perform routinely on spacecraft in demanding environments.

Concrete is also a composite material, and is used more than any other man-made material in the world. As of 2006, about 7.5 billion cubic metres of concrete are made each year—more than one cubic metre for every person on Earth.

Composition

Wood is a natural composite of Cellulose fibres in a matrix of lignin. The earliest man-made composite materials were straw and mud combined to form bricks for building construction. The ancient brick-making process can still be seen on Egyptian tomb paintings in the Metropolitan Museum of Art.

Composites are made up of individual materials referred to as constituent materials. There are two categories of constituent materials: matrix and reinforcement. At least one portion of each type is required. The matrix material surrounds and supports the reinforcement materials by maintaining their relative positions. The reinforcements impart their special mechanical and physical properties to enhance the matrix properties. A synergism produces material properties unavailable from the individual constituent materials, while the wide variety of matrix and strengthening materials allows the designer of the product or structure to choose an optimum combination.

Engineered composite materials must be formed to shape. The matrix material can be introduced to the reinforcement before or after the reinforcement material is placed into the mould cavity or onto the mould surface. The matrix material experiences a melding event, after which the part shape is essentially set. Depending upon the nature of the matrix material, this melding event can occur in various ways such as chemical polymerisation or solidification from the melted state.

A variety of moulding methods can be used according to the end-item design requirements. The principal factors impacting the methodology are the natures of the chosen matrix and reinforcement materials. Another important factor is the gross quantity of material to be produced. Large quantities can be used to justify high capital expenditures for rapid and automated manufacturing technology. Small production quantities are accommodated with lower capital expenditures but higher labour and tooling costs at a correspondingly slower rate.

Most commercially produced composites use a polymer matrix material often called a resin solution. There are many different polymers available depending upon the starting raw ingredients. There are several broad categories, each with numerous variations. The most common are known as polyester, vinyl ester, epoxy, phenolic, polyimide, polyamide, polypropylene, PEEK, and others. The reinforcement materials are often fibres but also commonly ground minerals. The various methods described below have been developed to reduce the resin content of the final product, or the fibre content is increased. As a rule of thumb, lay up results in a product containing 60% resin and 40% fibre, whereas vacuum infusion gives a final product with 40% resin and 60% fibre content. The strength of the product is greatly dependent on this ratio.

Moulding Methods

In general, the reinforcing and matrix materials are combined, compacted and processed to undergo a melding event. After the melding event, the part shape is essentially set, although it can deform under certain process conditions. For a thermoset polymeric matrix material, the melding event is a curing reaction that is initiated by the application of additional heat or chemical reactivity such as an organic peroxide. For a thermoplastic polymeric matrix material, the melding event is a solidification from the melted state. For a metal matrix material such as titanium foil, the melding event is a fusing at high pressure and a temperature near the melt point.

For many moulding methods, it is convenient to refer to one mould piece as a "lower" mould and another mould piece as an "upper" mould. Lower and upper refer to the different faces of the moulded panel, not the mould's configuration in space. In this convention, there is always a lower mould, and sometimes an upper mould. Part construction begins by applying materials to the lower mould. Lower mould and upper mould are more generalised descriptors than more common and specific terms such as male side, female side, a-side, b-side, tool side, bowl, hat, mandrel, etc. Continuous manufacturing processes use a different nomenclature.

The moulded product is often referred to as a panel. For certain geometries and material combinations, it can be referred to as a casting. For certain continuous processes, it can be referred to as a profile. Applied with a pressure roller, a spray device or manually. This process is generally done at ambient temperature and atmospheric pressure. Two variations of open moulding are Hand Layup and Spray-up.

Vacuum Bag Moulding

A process using a two-sided mould set that shapes both surfaces of the panel. On the lower side is a rigid mould and on the upper side is a flexible membrane or vacuum bag. The flexible membrane can be a reusable silicone material or an extruded polymer film. Then, vacuum is applied to the mould cavity. This process can be performed at either ambient or elevated temperature with ambient atmospheric pressure acting upon the vacuum bag. Most economical way is using a venturi vacuum and air compressor or a vacuum pump.

A vacuum bag is a bag made of strong rubber-coated fabric or a polymer film used to bond or laminate materials. In some applications the bag encloses the entire material, or in other applications a mold is used to form one face of the laminate with the bag being single sided to seal the outer face of the laminate to the mold. The open end is sealed and the air is drawn out of the bag through a nipple using a vacuum pump. As a result, uniform pressure approaching one atmosphere is applied to the surfaces of the object inside the bag, holding parts together while the adhesive cures. The entire bag may be placed in a temperature-controlled oven, oil bath or water bath and gently heated to accelerate curing.

In commercial woodworking facilities vacuum bags are used to laminate curved and irregular shaped workpieces.

Vacuum bagging is widely used in the composites industry as well. Carbon fibre fabric and fibreglass, along with resins and epoxies are common materials laminated together with a vacuum bag operation.

Typically, polyurethane or vinyl materials are used to make the bag, which is commonly open at both ends. This gives access to the piece, or pieces to be glued. A plastic rod is laid onto the bag, which is then folded over the rod. A plastic sleeve with an opening in it, is then snapped over the rod. This procedure forms a seal at both ends of the bag, when the vacuum is ready to be drawn.

A "platen" is used inside the bag for the piece being glued to lay on. The platen has a series of small slots cut into it, to allow the air under it to be evacuated. The platen must have rounded edges and corners to prevent the vacuum from tearing the bag.

When a curved part is to be glued in a vacuum bag, it is important that the pieces being glued be placed over a solidly built form, or have an air bladder placed under the form. This air bladder has access to "free air" outside the bag. It is used to create an equal pressure under the form, preventing it from being crushed.

Pressure Bag Moulding

This process is related to vacuum bag moulding in exactly the same way as it sounds. A solid female mould is used along with a flexible male mould. The reinforcement is placed inside the female mould with just enough resin to allow the fabric to stick in place (wet lay up). A measured amount of resin is then liberally brushed indiscriminately into the mould and the mould is then clamped to a machine that contains the male flexible mould. The flexible male membrane is then inflated with heated compressed air or possibly steam. The female mould can also be heated. Excess resin is forced out along with trapped air. This process is extensively used in the production of composite helmets due to the lower cost of unskilled labour. Cycle times for a helmet bag moulding machine vary from 20 to 45 minutes, but the finished shells require no further curing if the moulds are heated.

Autoclave Moulding

A process using a two-sided mould set that forms both surfaces of the panel. On the lower side is a rigid mould and on the upper side is a flexible membrane made from silicone or an extruded polymer film such as nylon. Reinforcement materials can be placed manually or robotically. They include continuous fibre forms fashioned into textile constructions. Most often, they are pre-impregnated with the resin in the form of prepreg fabrics or unidirectional tapes. In some instances, a resin film is placed upon the lower mould and dry reinforcement is placed above. The upper mould is installed and vacuum is applied to the mould cavity. The assembly is placed into an autoclave. This process is generally performed at both elevated pressure and elevated temperature. The use of elevated pressure facilitates a high fibre volume fraction and low void content for maximum structural efficiency.

Resin Transfer Moulding (RTM)

A process using a two-sided mould set that forms both surfaces of the panel. The lower side is a rigid mould. The upper side can be a rigid or flexible mould. Flexible moulds can be made from composite materials, silicone or extruded polymer films such as nylon. The two sides fit together to produce a mould cavity. The distinguishing feature of resin transfer moulding is that the reinforcement materials are placed into this cavity and the mould set is closed prior to the introduction of matrix material. Resin transfer moulding includes numerous varieties which differ in the mechanics of how the resin is introduced to the reinforcement in the mould cavity. These variations include everything from vacuum infusion to vacuum assisted resin transfer moulding (VARTM). This process can be performed at either ambient or elevated temperature.

Other

Other types of moulding include press moulding, transfer moulding, pultrusion moulding, filament winding, casting, centrifugal casting and continuous casting. There are also forming capabilities including CNC filament winding, vacuum infusion, wet lay-up, compression moulding, and thermoplastic moulding, to name a few. The use of curing ovens and paint booths is also needed for some projects.

Tooling

Some types of tooling materials used in the manufacturing of composites structures include invar, steel, aluminium, reinforced silicone rubber, nickel, and carbon fibre. Selection of the tooling material is typically based on, but not limited to, the coefficient of thermal expansion, expected number of cycles, end item tolerance, desired or required surface condition, method of cure, glass transition temperature of the material being moulded, moulding method, matrix, cost and a variety of other considerations.

Properties

Mechanics: The physical properties of composite materials are generally not isotropic (independent of direction of applied force) in nature, but rather are typically anisotropc (different depending on the direction of the applied force or load). For instance, the stiffness of a composite panel will often depend upon the orientation of the applied forces and/or moments. Panel stiffness is also dependent on the design of the panel. For instance, the fibre reinforcement and matrix used, the method of panel build, thermoset versus thermoplastic, type of weave, and orientation of fibre axis to the primary force.

In contrast, isotropic materials (for example, aluminium or steel), in standard wrought forms, typically have the same stiffness regardless of the directional orientation of the applied forces and/or moments.

The relationship between forces/moments and strains/curvatures for an isotropic material can be described with the following material properties: Young's Modulus, the shear Modulus and the Poisson's ratio, in relatively simple mathematical relationships. For the anisotropic material, it requires the mathematics of a second order tensor and up to 21 material property constants. For the special case of orthogonal isotropy, there are three different material property constants for each of Young's Modulus, Shear Modulus and Poisson's ratio—a total of 9 constants to describe the relationship between forces/moments and strains/curvatures. Techniques that take advantage of the anisotropic properties of the materials include mortise and tenon joints (in natural composites such as wood) and Pi Joints in synthetic composites.

Resins

Typically, most common composite materials, including fibreglass, carbon fibre, and Kevlar, include at least two parts, the substrate and the resin. Polyester resin tends to have yellowish tint, and is suitable for most backyard projects. Its weaknesses are that it is UV sensitive and can tend to degrade over time, and thus generally is also coated to help preserve it. It is often used in the making of surfboards and for marine applications. Its hardener is a MEKP, and is mixed at 14 drops per oz. MEKP is composed of methyl ethyl ketone peroxide, a catalyst. When MEKP is mixed with the resin, the resulting chemical reaction causes heat to build up and cure or harden the resin.

Vinylester resin tends to have a purplish to bluish to greenish tint. This resin has lower viscosity than polyester resin, and is more transparent. This resin is often billed as being fuel resistant, but will melt in contact with gasoline. This resin tends to be more resistant over time to degradation than polyester resin, and is more flexible. It uses the same hardener as polyester resin (at the same mix ratio) and the cost is approximately the same. Epoxy resin is almost totally transparent when cured. In the aerospace industry, epoxy is used as a structural matrix material or as a structural glue.

Shape memory polymer (SMP) resins have varying visual characteristics depending on their formulation. These resins may be epoxy-based, which can be used for auto body and outdoor equipment repairs; cyanate-ester-based, which are used in space applications; and acrylate-based, which can be used in very cold temperature applications, such as for sensors that indicate whether perishable goods have warmed above a certain maximum temperature. These resins are unique in that their shape can be repeatedly changed by heating above their glass transition temperature (Tg). When heated, they become flexible and elastic, allowing for easy configuration. Once they are cooled, they will maintain their new shape. The resins will return to their original shapes when they are reheated above their Tg. The advantage of shape memory polymer resins is that they can be shaped and reshaped repeatedly without losing their material properties, and these resins can be used in fabricating shape memory composites.

3

Quantum Statistical Mechanics

Quantum statistical mechanics is the study of statistical ensembles of quantum mechanical systems. A statistical ensemble is described by a density operator S, which is a non-negative, self-adjoint, trace-class operator of trace 1 on the Hilbert space H describing the quantum system. This can be shown under various mathematical formalisms for quantum mechanics. One such formalism is provided by quantum logic.

Expectation

From classical probability theory, we know that the expectation of a random variable X is completely determined by its distribution D_X by

$$\mathrm{Exp}(X) = \int_{\mathbb{R}} \lambda d\mathrm{D}_X(\lambda)$$

assuming, of course, that the random variable is integrable or that the random variable is non-negative. Similarly, let A be an observable of a quantum mechanical system. A is given by a densely defined self-adjoint operator on H. The spectral measure of A defined by

$$\mathrm{E}_A(U) = \int_U \lambda d\mathrm{E}(\lambda),$$

uniquely determines A and conversely, is uniquely determined by A. E_A is a boolean homomorphism from the Borel subsets of R into the lattice Q of self-adjoint projections of H. In analogy with probability theory, given a state S, we introduce the *distribution* of A under S which is the probability measure defined on the Borel subsets of R by

$$\mathrm{D}_A(U) = \mathrm{Tr}(\mathrm{E}_A(U)S).$$

Similarly, the expected value of A is defined in terms of the probability distribution D_A by

$$\text{Exp}(A) = \int_{\mathbb{R}} \lambda d\text{D}_A(\lambda).$$

Note that this expectation is relative to the mixed state S which is used in the definition of D_A.

Von Neumann Entropy

Of particular significance for describing randomness of a state is the von Neumann entropy of S *formally* defined by

$$\text{H}(S) = -\text{Tr}(S \log_2 S).$$

Actually, the operator $S \log_2 S$ is not necessarily trace-class. However, if S is a non-negative self-adjoint operator not of trace class we define Tr(S) = +–. Also note that any density operator S can be diagonalised, that it can be represented in some orthonormal basis by a (possibly infinite) matrix of the form

$$\begin{bmatrix} \lambda_1 & 0 & \cdots & 0 & \cdots \\ 0 & \lambda_2 & \cdots & 0 & \cdots \\ & & \cdots & & \\ 0 & 0 & \cdots & \lambda_n & \cdots \\ & & \cdots & \cdots & \end{bmatrix}$$

and we define

$$\text{H}(S) = -\sum_i \lambda_i \log_2 \lambda_i.$$

The convention is that $0 \log_2 0 = 0$, since an event with probability zero should not contribute to the entropy. This value is an extended real number (that is in [0, –]) and this is clearly a unitary invariant of S.

Remark. It is indeed possible that H(S) = +– for some density operator S. In fact T be the diagonal matrix

$$T = \begin{bmatrix} \frac{1}{2(\log_2 2)^2} & 0 & \cdots & 0 & \cdots \\ 0 & \frac{1}{3(\log_2 3)^2} & \cdots & 0 & \cdots \\ & & \cdots & & \\ 0 & 0 & \cdots & \frac{1}{n(\log_2 n)^2} & \cdots \\ & & \cdots & \cdots & \end{bmatrix}$$

T is non-negative trace class and one can show $T \log_2 T$ is not trace-class.

Theorem. Entropy is a unitary invariant.

In analogy with classical entropy (notice the similarity in the definitions), H(S) measures the amount of randomness in the state S. The more dispersed the eigenvalues are, the larger the system entropy. For a system in which the space H is finite-dimensional, entropy is maximised for the states S which in diagonal form have the representation

$$\begin{bmatrix} \frac{1}{n} & 0 & \cdots & 0 \\ 0 & \frac{1}{n} & \cdots & 0 \\ & & \cdots & \\ 0 & 0 & \cdots & \frac{1}{n} \end{bmatrix}$$

For such an S, H(S) = $\log_2 n$. The state S is called the maximally mixed state.

Recall that a pure state is one of the form

$$S = |\psi\rangle\langle\psi|,$$

for ψ a vector of norm 1.

Theorem. H(S) = 0 if and only if S is a pure state.

For S is a pure state if and only if its diagonal form has exactly one non-zero entry which is a 1.

Entropy can be used as a measure of quantum entanglement.

Gibbs Canonical Ensemble

Consider an ensemble of systems described by a Hamiltonian H with average energy E. If H has pure-point spectrum and the eigenvalues E_n of H go to + – sufficiently fast, $e^{-r H}$ will be a non-negative trace-class operator for every positive r.

The *Gibbs canonical ensemble* is described by the state

$$S = \frac{e^{-\beta H}}{\mathrm{Tr}(e^{-\beta H})}.$$

Where β is such that the ensemble average of energy satisfies

$$\mathrm{Tr}(SH) = E$$

and

$$\mathrm{Tr}(e^{-\beta H}) = \sum_n e^{-\beta E_n}$$

is the quantum mechanical version of the canonical partition function. The probability that a system chosen at random from the ensemble will be in a state corresponding to energy eigenvalue E_m is

$$\frac{e^{-\beta E_m}}{\sum_n e^{-\beta E_n}}.$$

Under certain conditions, the Gibbs canonical ensemble maximises the von Neumann entropy of the state subject to the energy conservation requirement.

Strength of Materials

In materials science, the strength of a material is its ability to withstand an applied stress without failure. The applied stress may be tensile, compressive, or shear. Strength of materials is a subject which deals with loads, deformations and the forces acting on a material. A load applied to a mechanical member will induce internal forces within the member called stresses. The stresses acting on the material cause deformation of the material. Deformation of the material is called strain, while the intensity of the internal forces is called stress. The strength of any material relies on three different type of analytical method: strength, stiffness and stability, where strength refers to the load carrying capacity, stiffness refers to the deformation or elongation, and stability refers to the ability to maintain its initial configuration. Material yield strength refers to the point on the engineering stress-strain curve (as opposed to true stress-strain curve) beyond which the material experiences deformations that will not be completely reversed upon removal of the loading. The ultimate strength refers to the point on the engineering stress-strain curve corresponding to the stress that produces fracture.

A material's strength is dependent on its microstructure. The engineering processes to which a material is subjected can alter this microstructure. The variety of strengthening mechanisms that alter the strength of a material includes work hardening, solid solution strengthening, precipitation hardening and grain boundary strengthening and can be quantitatively and qualitatively explained.

Strengthening mechanisms are accompanied by the caveat that some mechanical properties of the material may degenerate in an attempt to make the material stronger. For example, in grain boundary strengthening, although yield strength is maximised with decreasing grain size, ultimately, very small grain sizes make the material brittle. In general, the yield strength of a material is an adequate indicator of the material's mechanical strength. Considered in tandem with the fact that the yield strength is the parameter that predicts plastic deformation in the material, one can make informed decisions on how to increase the strength of a material depending its microstructural properties and the desired end effect. Strength is expressed in terms of compressive strength, tensile strength, and shear strength, namely the limit states of compressive stress, tensile stress and shear stress, respectively. The effects of dynamic loading are probably the most important practical consideration of the strength of materials, especially the problem of fatigue. Repeated loading often initiates brittle cracks, which grow until failure occurs. The cracks always start at stress concentrations, especially changes in cross-section of the product, near holes and corners.

The study of the subject of strength of materials often refers to various methods of calculating stresses in structural members, such as beams, columns and shafts. The methods employed to predict the response of a structure under loading and its susceptibility to various failure modes may take into account various properties of the materials other than material (yield or ultimate) strength. For example failure in buckling is dependent on material stiffness (Young's Modulus).

Types of Loadings

- Transverse loading — Forces applied perpendicular to the longitudinal axis of a member. Transverse loading causes the member to bend and deflect from its original position, with internal tensile and compressive strains accompanying the change in curvature of the member. Transverse loading also induces shear forces that cause shear deformation of the material and increase the transverse deflection of the member.
- Axial loading — The applied forces are collinear with the longitudinal axes of the member. The forces cause the member to either stretch or shorten.
- Torsional loading — Twisting action caused by a pair of externally applied equal and oppositely directed force couples

acting on parallel planes or by a single external couple applied to a member that has one end fixed against rotation.

Definitions

Stress Terms: Uniaxial stress is expressed by;

$$\sigma = \frac{F}{A},$$

where F is the force [N] acting on an area A [m^2]. The area can be the undeformed area or the deformed area, depending on whether engineering stress or true stress is of interest.

- *Compressive stress* (or compression) is the stress state caused by an applied load that acts to reduce the length of the material (compression member) in the axis of the applied load, in other words the stress state caused by squeesing the material. A simple case of compression is the uniaxial compression induced by the action of opposite, pushing forces. Compressive strength for materials is generally higher than their tensile strength. However, structures loaded in compression are subject to additional failure modes dependent on geometry, such as buckling.
- *Tensile stress* is the stress state caused by an applied load that tends to elongate the material in the axis of the applied load, in other words the stress caused by *pulling* the material. The strength of structures of equal cross sectional area loaded in tension is independent of shape of the cross section. Materials loaded in tension are susceptible to stress concentrations such as material defects or abrupt changes in geometry. However, materials exhibiting ductile behaviour (most metals for example) can tolerate some defects while brittle materials (such as ceramics) can fail well below their ultimate material strength.
- *Shear stress* is the stress state caused by a pair of opposing forces acting along parallel lines of action through the material, in other words the stress caused by faces of the material *sliding* relative to one another. An example is cutting paper with scissors or stresses due to torsional loading.

Strength Terms

- *Yield strength* is the lowest stress that produces a permanent deformation in a material. In some materials, like aluminium alloys, the point of yielding is difficult to identify, thus it is usually defined as the stress required to cause 0.2% plastic strain. This is called a 0.2% proof stress.

- *Compressive strength* is a limit state of compressive stress that leads to failure in the manner of ductile failure (infinite theoretical yield) or brittle failure (rupture as the result of crack propagation, or sliding along a weak plane).
- *Tensile strength* or *ultimate tensile strength* is a limit state of tensile stress that leads to tensile failure in the manner of ductile failure (yield as the first stage of that failure, some hardening in the second stage and breakage after a possible "neck" formation) or brittle failure (sudden breaking in two or more pieces at a low stress state). Tensile strength can be quoted as either true stress or engineering stress.
- *Fatigue strength* is a measure of the strength of a material or a component under cyclic loading, and is usually more difficult to assess than the static strength measures. Fatigue strength is quoted as stress amplitude or stress range ($\Delta\sigma = \sigma_{max} - \sigma_{min}$), usually at zero mean stress, along with the number of cycles to failure under that condition of stress.
- *Impact strength*, is the capability of the material to withstand a suddenly applied load and is expressed in terms of energy. Often measured with the Izod impact strength test or Charpy impact test, both of which measure the impact energy required to fracture a sample. Volume, modulus of elasticity, distribution of forces, and yield strength effect the impact strength of a material. In order for a material or object to have a higher impact strength the stresses must be distributed evenly throughout the object. It also must have a large volume with a low modulus of elasticity and a high material yield strength.

Strain (Deformation) Terms

- *Deformation* of the material is the change in geometry created when stress is applied (in the form of force loading, gravitational field, acceleration, thermal expansion, etc.). Deformation is expressed by the displacement field of the material.
- *Strain* or *reduced deformation* is a mathematical term that expresses the trend of the deformation change among the material field. Strain is the deformation per unit length. In the case of uniaxial loading - displacements of a specimen (for example a bar element) strain is expressed as the quotient of the displacement and the length of the specimen. For 3D

displacement fields it is expressed as derivatives of displacement functions in terms of a second order tensor (with 6 independent elements).

- *Deflection* is a term to describe the magnitude to which a structural element bends under a load.

Stress-Strain Relations

- *Elasticity* is the ability of a material to return to its previous shape after stress is released. In many materials, the relation between applied stress is directly proportional to the resulting strain (up to a certain limit), and a graph representing those two quantities is a straight line.

The slope of this line is known as Young's Modulus, or the "Modulus of Elasticity." The Modulus of Elasticity can be used to determine the stress-strain relationship in the linear-elastic portion of the stress-strain curve. The linear-elastic region is either below the yield point, or if a yield point is not easily identified on the stress-strain plot it is defined to be between 0 and 0.2% strain, and is defined as the region of strain in which no yielding (permanent deformation) occurs.

- *Plasticity* or plastic deformation is the opposite of elastic deformation and is defined as unrecoverable strain. Plastic deformation is retained after the release of the applied stress. Most materials in the linear-elastic category are usually capable of plastic deformation. Brittle materials, like ceramics, do not experience any plastic deformation and will fracture under relatively low stress. Materials such as metals usually experience a small amount of plastic deformation before failure while ductile metals such as copper and lead or polymers will plasticly deform much more.

Consider the difference between a carrot and chewed bubble gum. The carrot will stretch very little before breaking. The chewed bubble gum, on the other hand, will plastically deform enormously before finally breaking.

Design Terms

Ultimate strength is an attribute related to a material, rather than just a specific specimen made of the material, and as such it is quoted as the force per unit of cross section area (N/m^2). The ultimate strength is the maximum stress that a material can withstand before it breaks or weakens. For example, the ultimate tensile strength (UTS) of AISI 1018 Steel is 440 MN/m^2. In general, the SI unit of stress is the pascal, where 1 Pa = 1 N/m^2. In Imperial units, the unit of stress is given as lbf/in^2 or pounds-force per square inch. This unit is often abbreviated as psi. One thousand psi is abbreviated ksi.

A Factor of safety is a design criteria that an engineered component or structure must achieve. *FS = UTS / R*, where FS: the factor of safety, R: The applied stress, and UTS: ultimate stress (psi or N/m^2)

Margin of Safety is also sometimes used to as design criteria. It is defined MS = Failure Load/(Factor of Safety * Predicted Load) - 1

For example to achieve a factor of safety of 4, the allowable stress in an AISI 1018 steel component can be calculated to be *R = UTS / FS* = 440/4 = 110 MPa, or $R = 110\times10^6$ N/m^2. Such allowable stresses are also known as "design stresses" or "working stresses."

Design stresses that have been determined from the ultimate or yield point values of the materials give safe and reliable results only for the case of static loading. Many machine parts fail when subjected to a non steady and continuously varying loads even though the developed stresses are below the yield point. Such failures are called fatigue failure. The failure is by a fracture that appears to be brittle with little or no visible evidence of yielding. However, when the stress is kept below "fatigue stress" or "endurance limit stress", the part will endure indefinitely. A purely reversing or cyclic stress is one that alternates between equal positive and negative peak stresses during each cycle of operation. In a purely cyclic stress, the average stress is zero. When a part is subjected to a cyclic stress, also known as stress range (Sr), it has been observed that the failure of the part occurs after a number of stress reversals (N) even if the magnitude of the stress range is below the material's yield strength. Generally, higher the range stress, the fewer the number of reversals needed for failure.

Failure Theories

There are four important failure theories, namely (1) maximum shear stress theory, (2) maximum normal stress theory, (3) maximum strain energy theory, and (4) maximum distortion energy theory. Out of these four theories of failure, the maximum normal stress theory is only applicable for brittle materials, and the remaining three theories are applicable for ductile materials.

- Maximum Shear stress Theory- This theory postulates that failure will occur in a machine part if the magnitude of the maximum shear stress (tmax) in the part exceeds the shear strength (typ) of the material determined from uniaxial testing.

This theory postulates, that failure will occur when, tmax = typ or max of [|S1-S2|/2 , |S2-S3|/2 , and |S3-S1|/2] = Syp/2 Dividing both side by 2, max of [|S1-S2| , |S2-S3|, and |S3-S1|] = Syp Using a design factor of safety Nfs, the theory formulates the design equation as, max of [|S1-S2| , |S2-S3|, and |S3-S1|] should be less than or equal to Syp/Nfs

- Maximum normal stress theory- this theory postulates, that failure will occur in machine part if the maximum normal stress in the part exceeds the ultimate tensile stress of the material as determined from uniaxial testing. This theory deals with brittle materials only. The maximum tensile stress should be less than or equal to ultimate tensile stress divided by factor of safety. The magnitude of the maximum compressive stress should be less than ultimate compressive stress divided by factor of safety.

As the three principal stresses at a point in the part S1, S2, or S3 may be both tensile and compressive stresses, when this theory is applied, we need to check for failures both from tension and compression. The method of application of this theory is to find the maximum tensile stress, and the maximum compressive stress from the given values of S1, S2, and S3. The largest positive value among S1, S2, and S3 is the maximum tensile stress and the smallest negative value is the maximum compressive stress. For example if S1 = 80 MPa, S2 = -100 MPa, and S3 = -150 MPa, then the maximum tensile stress = 80 MPa, and the maximum compressive stress = -150 MPa (smallest negative value!). Thus according to this theory, the safe design condition for brittle material can be given by: The maximum tensile stress should be less than or equal to Sut/Nfs and The magnitude of the maximum compressive stress should less than Suc/Nfs

- Maximum strain energy theory-this theory postulates that failure will occur when the strain energy per unit volume due to the applied stresses in a part equals the strain energy per unit volume at the yield point in uniaxial testing.

Strain energy is the energy stored in a material due elastic deformation, which is, work done during elastic deformation. Work done per unit volume = strain x average stress. During tensile test, stress increases from zero to Syp, that is average stress = Syp/2. Elastic strain at yield point = Syp/E, where E is the elastic modulus of elasticity. Strain energy per unit volume during uniaxial tension = average stress x strain = Syp2/2E

- Maximum distortion energy theory- this theory is also known as shear energy theory or von Mises-Hencky theory. This theory postulates that failure will occur when the distortion energy per unit volume due to the applied stresses in a part equals the distortion energy per unit volume at the yield point in uniaxial testing. The total elastic energy due to strain can be divided into two parts. One part causes change in volume, and the other part causes change in shape. Distortion energy is the amount of energy that is needed to change the shape.

Application of Failure Theory

Out of the four theories, only the maximum normal stress theory predicts failure for brittle materials. The rest of the three theories are applicable for ductile materials. Out of these three, the distortion energy theory provides most accurate results in majority of the stress conditions. The strain energy theory needs the value of Poisson's ratio of the part material, which is often not readily available. The maximum shear stress theory is conservative. For simple unidirectional normal stresses all theories are equivalent, which means all theories will give the same result.

Kinematics

Kinematics is the branch of classical mechanics that describes the motion of bodies (objects) and systems (groups of objects) without consideration of the forces that cause the motion.

Kinematics is not to be confused with another branch of classical mechanics: analytical dynamics (the study of the relationship between the motion of objects and its causes), sometimes subdivided into *kinetics* (the study of the relation between external forces and motion) and *statics* (the study of the relations in a system at equilibrium). *Kinematics* also differs from *dynamics* as used in modern-day physics to describe time-evolution of a system.

The term *kinematics* is less common today than in the past, but still has a role in physics. The term *kinematics* also finds use in biomechanics and animal locomotion. Further, mathematicians that include time as a parameter in geometry have developed the subject of kinematic geometry.

The simplest application of kinematics is for particle motion, translational or rotational. The next level of complexity comes from the introduction of rigid bodies, which are collections of particles having time invariant distances between themselves. Rigid bodies might undergo translation and rotation or a combination of both. A more complicated case is the kinematics of a *system* of rigid bodies, which may be linked together by mechanical joints. Kinematics can be used to find the possible range of motion for a given mechanism, or, working in reverse, can be used to design a mechanism that has a desired range of motion. The movement of a crane and the oscillations of a piston in an engine are both simple kinematic systems. The crane is a type of open kinematic chain, while the piston is part of a closed four-bar linkage.

Linear Motion

Linear or translational kinematics is the description of the motion in space of a point along a *line*, also known as a *trajectory* or *path*. This path can be either straight (*rectilinear*) or curved (*curvilinear*).

Particle Kinematics

Particle kinematics is the study of the kinematics of a single particle. The results obtained in particle kinematics are used to study the kinematics of collection of particles, dynamics and in many other branches of mechanics.

Position and Reference Frames

The position of a point in space is the most fundamental idea in particle kinematics. To specify the position of a point, one must specify three things: the reference point (often called the origin), distance from the reference point and the direction in space of the straight line from the reference point to the particle. Exclusion of any of these three parameters renders the description of position incomplete. Consider for example a tower 50 m south from your home. The reference point is home, the distance 50 m and the direction south. If one only says that the tower is 50 m south, the natural question that arises is "from where?" If one says that the tower is southward from your home, the question that arises is "how far?" If one says the tower is 50 m from your home, the question that arises is "in which direction?" Hence, all these three parameters are crucial to defining uniquely the position of a point in space.

Position is usually described by mathematical quantities that have all these three attributes: the most common are vectors and complex numbers. Usually, only vectors are used. For measurement of distances and directions, usually three dimensional coordinate systems are used with the origin coinciding with the reference point. A three-dimensional coordinate system (whose origin coincides with the reference point) with some provision for time measurement is called a reference frame or frame of reference or simply frame. All observations in physics are incomplete without the reference frame being specified.

Position Vector

The position vector of a particle is a vector drawn from the origin of the reference frame to the particle. It expresses both the distance of the point from the origin and its sense from the origin. In three dimensions, the position of point A can be expressed as

$$\mathrm{r}_A = (x_A, y_A, z_A),$$

where x_A, y_A, and z_A are the Cartesian coordinates of the point. The magnitude of the position vector $|\mathrm{r}|$ gives the distance between the point A and the origin.

$$|\mathrm{r}| = \sqrt{x_A^2 + y_A^2 + z_A^2}.$$

The direction cosines of the position vector provide a quantitative measure of direction. It is important to note that the position vector of a particle isn't unique. The position vector of a given particle is different relative to different frames of reference.

Rest and Motion

Once the concept of position is firmly established, the ideas of rest and motion naturally follow. If the position vector of the particle (relative to a given reference frame) changes with time, then the particle is said to be in motion with respect to the chosen reference frame. However, if the position vector of the particle (relative to a given reference frame) remains the same with time, then the particle is said to be at rest with respect to the chosen frame. Note that rest and motion are relative to the reference frame chosen. It is quite possible that a particle at rest relative to a particular reference frame is in motion relative to the other. Hence, rest and motion aren't absolute terms, rather they are dependent on reference frame. For example, a passenger in a moving car may be at rest with respect to the car, but in motion with respect to the road.

Path

A particle's path is the locus between its beginning and end points which is reference-frame dependent. The path of a particle may be rectilinear (straight line) in one frame, and curved in another.

Displacement

Displacement is a vector describing the difference in position between two points, i.e. it is the change in position the particle undergoes during the time interval. If point A has position $\mathrm{r}_A = (x_A, y_A, z_A)$ and point B has position $\mathrm{r}_B = (x_B, y_B, z_B)$, the displacement r_{AB} of B from A is given by

$$\mathrm{r}_{AB} = \mathrm{r}_B - \mathrm{r}_A = (x_B - x_A, y_B - y_A, z_B - z_A).$$

Geometrically, displacement is the shortest distance between the points A and B. Displacement, distinct from position vector, is independent of the reference frame. This can be understood as follows: the positions of points is frame dependent, however, the shortest distance between any pair of points is invariant on translation from one frame to another (barring relativistic cases).

Distance

Distance is a scalar quantity, describing the length of the path between two points along which a particle has travelled.

When considering the motion of a particle over time, distance is the length of the particle's path and may be different from displacement, which is the change from its initial position to its final position. For example, a race car traversing a 10 km closed loop from start to finish travels a distance of 10 km; its displacement, however, is zero because it arrives back at its initial position.

If the position of the particle is known as a function of time ($\mathrm{r} = \mathrm{r}(t)$), the distance s it travels from time t_1 to time t_2 can be found by

$$s = \int_{t_1}^{t_2} |d\mathrm{r}| = \int_{t_1}^{t_2} ds = \int_{t_1}^{t_2} \sqrt{dx^2 + dy^2 + dz^2} = \int_{t_1}^{t_2} \sqrt{\left(\frac{dx}{dt}\right)^2 + \left(\frac{dy}{dt}\right)^2 + \left(\frac{dz}{dt}\right)^2}\, dt.$$

The formula utilises the fact that over an infinitesimal time interval, the magnitude of the displacement equals the distance covered in that interval. This is analogous to the geometric fact that infinitesimal arcs on a curved line coincide with the chord drawn between the ends of the arc itself.

Velocity and Speed

Average velocity is defined as

$$\bar{\mathrm{v}} = \frac{\Delta \mathrm{r}}{\Delta t},$$

where Δr is the change in position and Δt is the interval of time over which the position changes. The direction of v is same as the direction of the change in position Δr as Δt>0.

Velocity is the measure of the rate of change in position with respect to time, that is, how the distance of a point changes with each instant of time. Velocity also is a vector. Instantaneous velocity (the velocity at an instant of time) can be defined as the limiting value of average velocity as the time interval Δt becomes smaller and smaller. Both Δr and Δt approach zero but the ratio v approaches a non-zero limit v. That is,

$$v = \lim_{\Delta t \to 0} \frac{\Delta r}{\Delta t} = \frac{dr}{dt},$$

where dr is an infinitesimally small displacement and dt is an infinitesimally small length of time. As per its definition in the derivative form, velocity can be said to be the time rate of change of position. Further, as dr is tangential to the actual path, so is the velocity.

As a position vector itself is frame dependent, velocity is also dependent on the reference frame.

The speed of an object is the magnitude $|v|$ of its velocity. It is a scalar quantity:

$$v = \lim_{\Delta t \to 0} \frac{\Delta r}{\Delta t} = \frac{dr}{dt},$$

The distance travelled by a particle over time is a non-decreasing quantity. Hence, ds/dt is non-negative, which implies that speed is also non-negative.

Acceleration

Average acceleration (acceleration over a length of time) is defined as:

$$\bar{a} = \frac{\Delta v}{\Delta t},$$

where Δv is the change in velocity and Δt is the interval of time over which velocity changes.

Acceleration is the vector quantity describing the rate of change with time of velocity. Instantaneous acceleration (the acceleration at an instant of time) is defined as the limiting value of average acceleration as Δt becomes smaller and smaller. Under such a limit, $a \to a$.

$$a = \lim_{\Delta t \to 0} \frac{\Delta v}{\Delta t} = \frac{dv}{dt},$$

where dv is an infinitesimally small change in velocity and dt is an infinitesimally small length of time.

Types of Motion Based on Velocity and Acceleration

If the acceleration of a particle is zero, then the velocity of the particle is constant over time and the motion is said to be *uniform*. Otherwise, the motion is *non-uniform*.

If the acceleration is non-zero but constant, the motion is said to be *motion with constant acceleration*. On the other hand, if the acceleration is variable, the motion is called *motion with variable acceleration*. In motion with variable acceleration, the rate of change of acceleration is called the *jerk*.

Integral relations

The above definitions can be inverted by mathematical integration to find:

$$v(t) = v_0 + \int_{t_0}^{t} a(t)\, dt$$

$$\begin{aligned} r(t) &= r_0 + \int_{t_0}^{t} v(t)\, dt \\ &= r_0 + v_0 t + \int_{t_0}^{t} \left[\int_{t_0}^{t} a(t) dt \right] dt \end{aligned}$$

Kinematics of Constant Acceleration

Many physical situations can be modelled as constant-acceleration processes, such as projectile motion.

Integrating acceleration a with respect to time t gives the change in velocity. When acceleration is constant both in direction and in magnitude, the point is said to be undergoing *uniformly accelerated motion*. In this case, the integral relations can be simplified:

$$v(t) = \int_0^t a\, dt' = v_0 + at.$$

$$\begin{aligned} r(t) &= r_0 + \int_0^t v\, dt' = r_0 + \int_0^t (v_0 + at)\, dt' \\ &= r_0 + v_0 t + \tfrac{1}{2} a t^2. \end{aligned}$$

Additional relations between displacement, velocity, acceleration, and time can be derived. Since $a = (v - v_0)/t$,

$$r(t) = r_0 + \left(\frac{v + v_0}{2}\right)t.$$

By using the definition of an average, this equation states that when the acceleration is constant average velocity times time equals displacement.

A relationship without explicit time dependence may also be derived for one-dimensional motion. Noting that $at = v - v_0$,

$$(r - r_0)\cdot at = (v - v_0)\cdot\frac{v + v_0}{2}t,$$

where $\cdot$ denotes the dot product. Dividing the t on both sides and carrying out the dot-products:

$$2(r - r_0)\cdot a = v^2 - v_0^2.$$

In the case of straight-line motion, $(r - r_0)$ is parallel to a. Then

$$v^2 = v_0^2 + 2a(r - r_0).$$

This relation is useful when time is not known explicitly.

Relative Velocity

To describe the motion of object *A* with respect to object *B*, when we know how each is moving with respect to a reference object *O*, we can use vector algebra. Choose an origin for reference, and let the positions of objects *A*, *B*, and *O* be denoted by r_A, r_B, and r_O. Then the position of *A* relative to the reference object *O* is

$$r_{A/O} = r_A - r_O$$

Consequently, the position of *A* relative to *B* is

$$r_{A/B} = r_A - r_B = r_A - r_O - (r_B - r_O) = r_{A/O} - r_{B/O}.$$

The above relative equation states that the motion of A relative to B is equal to the motion of A relative to O minus the motion of B relative to O. It may be easier to visualise this result if the terms are re-arranged:

$$r_{A/O} = r_{A/B} + r_{B/O},$$

or, in words, the motion of *A* relative to the reference is that of *B* plus the relative motion of *A* with respect to *B*. These relations between displacements become relations between velocities by simple time-differentiation, and a second differentiation makes them apply to accelerations.

For example, let Ann move with velocity V relative to the reference (we drop the *O* subscript for convenience) and let Bob move with velocity V_B, each velocity given with respect to the ground (point *O*). To find how fast Ann is moving relative to Bob (we call this velocity $V_{A/B}$), the equation above gives:

$$V_A = V_B + V_{A/B}.$$

To find $V_{A/B}$ we simply rearrange this equation to obtain:

$$V_{A/B} = V_A - V_B.$$

At velocities comparable to the speed of light, these equations are not valid. They are replaced by equations derived from Einstein's theory of special relativity.

Kinematics is the study of how things move. Here, we are interested in the motion of normal objects in our world. A normal object is visible, has edges, and has a location that can be expressed with (x, y, z) coordinates. We will not be discussing the motion of atomic particles or black holes or light.

We will create a vocabulary and a group of mathematical methods that will describe this ordinary motion. Understand that we will be developing a language for describing motion only. We won't be concerned with what is causing or changing the motion, or more correctly, the momentums of the objects. In other words, we are not concerned with the action of forces within this topic.

Rotational Motion

Rotational or angular kinematics is the description of the rotation of an object. The description of rotation requires some method for describing orientation, for example, the Euler angles. In what follows, attention is restricted to simple rotation about an axis of fixed orientation. The *z*-axis has been chosen for convenience.

Description of rotation then involves these three quantities:

- *Angular Position:* The oriented distance from a selected origin on the rotational axis to a point of an object is a vector r (*t*) locating the point. The vector r(*t*) has some projection (or, equivalently, some component) $r_\perp (t)$ on a plane perpendicular to the axis of rotation. Then the *angular position* of that point is the angle θ from a reference axis (typically the positive *x*-axis) to the vector $r_\perp (t)$ in a known rotation sense (typically given by the right-hand rule).

- *Angular Velocity:* The angular velocity ω is the rate at which the angular position θ changes with respect to time t:

$$\omega = \frac{d\theta}{dt}$$

The angular velocity is represented by a vector Ω pointing along the axis of rotation with magnitude ω and sense determined by the direction of rotation as given by the right-hand rule.

- *Angular Acceleration:* The magnitude of the angular acceleration α is the rate at which the angular velocity ω changes with respect to time t:

$$\alpha = \frac{d\omega}{dt}$$

The equations of translational kinematics can easily be extended to planar rotational kinematics with simple variable exchanges:

$$\omega_f = \omega_i + \alpha t$$

$$\theta_f - \theta_i = \omega_i t + \tfrac{1}{2}\alpha t^2$$

$$\theta_f - \theta_i = \tfrac{1}{2}(\omega_f + \omega_i)t$$

$$\omega_f^2 = \omega_i^2 + 2\alpha(\theta_f - \theta_i).$$

Here θ_i and θ_f are, respectively, the initial and final angular positions, ω_i and ω_f are, respectively, the initial and final angular velocities, and α is the constant angular acceleration. Although position in space and velocity in space are both true vectors (in terms of their properties under rotation), as is angular velocity, angle itself is not a true vector.

Point Object in Circular Motion

This example deals with a "point" object, by which is meant that complications due to rotation of the body itself about its own center of mass are ignored.

Displacement. An object in circular motion is located at a position r(t) given by:

$$r(t) = R u_R(t),$$

where u_R is a unit vector pointing outward from the axis of rotation towards the periphery of the circle of motion, located at a radius R from the axis.

Linear velocity. The velocity of the object is then

$$v(t) = \frac{d}{dt} r(t) = R \frac{d}{dt} u_R(t).$$

The magnitude of the unit vector u_R (by definition) is fixed, so its time dependence is entirely due to its rotation with the radius to the object, that is,

$$\frac{d}{dt} u_R(t) = \Omega \times u_R = \omega(t) u_\theta,$$

where u_θ is a unit vector perpendicular to u_R pointing in the direction of rotation, $\omega(t)$ is the (possibly time varying) angular rate of rotation, and the symbol × denotes the vector cross product. The velocity is then:

$$v(t) = R\omega(t) u_\theta.$$

The velocity therefore is tangential to the circular orbit of the object, pointing in the direction of rotation, and increasing in time if ω increases in time.

Linear acceleration. In the same manner, the acceleration of the object is defined as:

$$\begin{aligned} a(t) &= \frac{dv}{dt} = R \frac{d}{dt}(\omega u_\theta) \\ &= u_\theta R \frac{d\omega}{dt} + R\omega \frac{du_\theta}{dt} \\ &= u_\theta R \frac{d\omega}{dt} + R\omega \Omega \times u_\theta \\ &= u_\theta R \frac{d\omega}{dt} - u_R \omega^2 R \\ &= a_\theta(t) + a_R(t), \end{aligned}$$

which shows a leading term a_θ in the acceleration tangential to the orbit related to the angular acceleration of the object (supposing ω to vary in time) and a second term a_R directed inward from the object towards the center of rotation, called the centripetal acceleration.

Coordinate Systems

In any given situation, the most useful coordinates may be determined by constraints on the motion, or by the geometrical nature of the force causing or affecting the motion. Thus, to describe the motion of a bead constrained to move along a circular hoop, the most useful coordinate may be its angle on the hoop.

Similarly, to describe the motion of a particle acted upon by a central force, the most useful coordinates may be polar coordinates. Polar coordinates are extended into three dimensions with either the spherical polar or cylindrical polar coordinate systems. These are most useful in systems exhibiting spherical or cylindrical symmetry respectively.

Fixed Rectangular Coordinates

In this coordinate system, vectors are expressed as an addition of vectors in the x, y, and z direction from a non-rotating origin. Usually i, j, k are unit vectors in the *x*-, *y*-, and *z*-directions.

The position vector, r, the velocity vector, v, and the acceleration vector, a are expressed using rectangular coordinates in the following way:

$$\mathrm{r} = x\hat{\mathrm{i}} + y\hat{\mathrm{j}} + z\hat{\mathrm{k}}$$

$$\mathrm{v} = \dot{\mathrm{r}} = \dot{x}\hat{\mathrm{i}} + \dot{y}\hat{\mathrm{j}} + \dot{z}\hat{\mathrm{k}}$$

$$\mathrm{a} = \ddot{\mathrm{r}} = \ddot{x}\hat{\mathrm{i}} + \ddot{y}\hat{\mathrm{j}} + \ddot{z}\hat{\mathrm{k}}$$

Note: $\dot{x} = \dfrac{dx}{dt}, \quad \ddot{x} = \dfrac{d^2x}{dt^2}$

Two-Dimensional Rotating Reference Frame

This coordinate system expresses only planar motion. It is based on three orthogonal unit vectors: the vector i, and the vector j which form a basis for the plane in which the objects we are considering reside, and k about which rotation occurs. Unlike rectangular coordinates, which are measured relative to an origin that is fixed and non-rotating, the origin of these coordinates can rotate and translate - often following a particle on a body that is being studied.

Derivatives of Unit Vectors

The position, velocity, and acceleration vectors of a given point can be expressed using these coordinate systems, but we have to be a bit more careful than we do with fixed frames of reference. Since the frame of reference is rotating, the unit vectors also rotate, and this rotation must be taken into account when taking the derivative of any of these vectors. If the coordinate frame is rotating at angular rate ω in the counterclockwise direction (that is, $\Omega = \omega$ k using the right hand rule) then the derivatives of the unit vectors are as follows:

$$\dot{\hat{i}} = \omega\hat{k} \times \hat{i} = \omega\hat{j}$$

$$\dot{\hat{j}} = \omega\hat{k} \times \hat{j} = -\omega\hat{i}$$

Position, Velocity, and Acceleration

Given these identities, we can now figure out how to represent the position, velocity, and acceleration vectors of a particle using this reference frame.

Kinematic Constraints

A kinematic constraint is any condition relating properties of a dynamic system that must hold true at all times. Below are some common examples:

Rolling without Slipping

An object that rolls against a surface without slipping obeys the condition that the velocity of its center of mass is equal to the cross product of its angular velocity with a vector from the point of contact to the center of mass,

$$r = x\hat{i} + y\hat{j}.$$

For the case of an object that does not tip or turn, this reduces to v = R ω.

Inextensible Cord

This is the case where bodies are connected by an idealised cord that remains in tension and cannot change length. The constraint is that the sum of lengths of all segments of the cord is the total length, and accordingly the time derivative of this sum is zero. A dynamic problem of this type is the pendulum. Another example is a drum turned by the pull of gravity upon a falling weight attached to the rim by the inextensible cord. An *equilibrium* problem (not kinematic) of this type is the catenary.

Dynamics (Mechanics)

In the field of physics, the study of the causes of motion and changes in motion is dynamics. In other words the study of forces and why objects are in motion. *Dynamics* includes the study of the effect of torques on motion. These are in contrast to kinematics, the branch of classical mechanics that describes the motion of objects without consideration of the causes leading to the motion.

Generally speaking, researchers involved in dynamics study how a physical system might develop or alter over time and study the causes of those changes. In addition, Isaac Newton established the undergirding physical laws which govern dynamics in physics. By studying his system of mechanics, dynamics can be understood. In particular dynamics is mostly related to Newton's second law of motion. However, all three laws of motion are taken into consideration, because these are interrelated in any given observation or experiment.

For classical electromagnetism, it is Maxwell's equations that describe the dynamics. And the dynamics of classical systems involving both mechanics and electromagnetism are described by the combination of Newton's laws, Maxwell's equations, and the Lorentz force.

Force

From Newton, force can be defined as an exertion or pressure which can cause an object to move. The concept of force is used to describe an influence which causes a free body (object) to accelerate. It can be a push or a pull, which causes an object to change direction, have new velocity, or to deform temporarily or permanently. Generally speaking, force causes an object's state of motion to change.

Newton's Laws

Newton described force as the ability to cause a mass to accelerate.

- Newton's first law states that an object in motion will stay in motion unless a force is applied. This law deals with inertia, which is a property of matter that resists acceleration and depends only on mass.
- Newton's second law states that force quantity is equal to mass multiplied by the acceleration (F = ma).
- Newton's third law states that for every action, there is an equal but opposite reaction.

Robotics

Robotics is the branch of technology that deals with the design, construction, operation, structural disposition, manufacture and application of robots. Robotics is related to the sciences of electronics, engineering, mechanics, and software.

Etymology

The word *robotics* was derived from the word *robot*, which was introduced to the public by Czech writer Karel Èapek in his play *R.U.R. (Rossum's Universal Robots)*, which premiered in 1921.

According to the *Oxford English Dictionary,* the word *robotics* was first used in print by Isaac Asimov, in his science fiction short story "Liar!", published in May 1941 in *Astounding Science Fiction.* Asimov was unaware that he was coining the term; since the science and technology of electrical devices is *electronics*, he assumed *robotics* already referred to the science and technology of robots. In some of Asimov's other works, he states that the first use of the word *robotics* was in his short story *Runaround* (Astounding Science Fiction, March 1942). However, the word *robotics* appears in "Liar!"

History

The word *robot* was introduced to the public by the Czech writer Karel Èapek in his play *R.U.R. (Rossum's Universal Robots),* published in 1920. The play begins in a factory that makes artificial people called *robots* creatures who can be mistaken for humans – though they are closer to the modern ideas of androids. Karel Èapek himself did not coin the word. He wrote a short letter in reference to an etymology in the *Oxford English Dictionary* in which he named his brother Josef Èapek as its actual originator.

In 1927 the *Maschinenmensch* ("machine-human") gynoid humanoid robot (also called "Parody", "Futura", "Robotrix", or the "Maria impersonator") was the first and perhaps the most memorable depiction of a robot ever to appear on film was played by German actress Brigitte Helm) in Fritz Lang's film Metropolis.

In 1942 the science fiction writer Isaac Asimov formulated his Three Laws of Robotics and, in the process of doing so, coined the word "robotics".

In 1948 Norbert Wiener formulated the principles of cybernetics, the basis of practical robotics.

Fully autonomous robots only appeared in the second half of the 20th century. The first digitally operated and programmable robot, the Unimate, was installed in 1961 to lift hot pieces of metal from a die casting machine and stack them. Commercial and industrial robots are widespread today and used to perform jobs more cheaply, or more accurately and reliably, than humans. They are also employed in jobs which are too dirty, dangerous, or dull to be suitable for humans. Robots are widely used in manufacturing, assembly, packing and packaging, transport, earth and space exploration, surgery, weaponry, laboratory research, safety, and the mass production of consumer and industrial goods.

Power Source

At present; mostly (lead-acid) batteries are used, but potential power sources could be:

- pneumatic (compressed gases)
- hydraulics (liquids)
- flywheel energy storage
- organic garbage (through anaerobic digestion)
- faeces (human, animal); may be interesting in a military context as faeces of small combat groups may be reused for the energy requirements of the robot assistant
- still unproven energy sources: for example Nuclear fusion, as yet not used in nuclear reactors whereas Nuclear fission is proven (although there are not many robots using it as a power source apart from the Chinese rover tests.).
- radioactive source (such as with the proposed Ford car of the '50s); to those proposed in movies such as *Red Planet.*

Actuation

Actuators are like the "muscles" of a robot, the parts which convert stored energy into movement. By far the most popular actuators are electric motors that spin a wheel or gear, and linear actuators that control industrial robots in factories. But there are some recent advances in alternative types of actuators, powered by electricity, chemicals, or compressed air:

Electric Motors

The vast majority of robots use electric motors, often brushed and brushless DC motors in portable robots or AC motors in industrial robots and CNC machines.

Linear Actuators

Various types of linear actuators move in and out instead of by spinning, particularly when very large forces are needed such as with industrial robotics. They are typically powered by compressed air (pneumatic actuator) or an oil (hydraulic actuator).

Series Elastic Actuators

A spring can be designed as part of the motor actuator, to allow improved force control. It has been used in various robots, particularly walking humanoid robots.

Air Muscles

Pneumatic artificial muscles, also known as air muscles, are special tubes that contract (typically up to 40%) when air is forced inside it. They have been used for some robot applications.

Muscle Wire

Muscle wire, also known as Shape Memory Alloy, Nitinol or Flexinol Wire, is a material that contracts slightly (typically under 5%) when electricity runs through it. They have been used for some small robot applications.

Electroactive Polymers

EAPs or EPAMs are a new plastic material that can contract substantially (up to 400%) from electricity, and have been used in facial muscles and arms of humanoid robots, and to allow new robots to float, fly, swim or walk.

Piezo Motors

A recent alternative to DC motors are piezo motors or ultrasonic motors. These work on a fundamentally different principle, whereby tiny piezoceramic elements, vibrating many thousands of times per second, cause linear or rotary motion.

There are different mechanisms of operation; one type uses the vibration of the piezo elements to walk the motor in a circle or a straight line. Another type uses the piezo elements to cause a nut to vibrate and drive a screw. The advantages of these motors are nanometer resolution, speed, and available force for their size. These motors are already available commercially, and being used on some robots.

Elastic Nanotubes

Elastic nanotubes are a promising artificial muscle technology in early-stage experimental development. The absence of defects in carbon nanotubes enables these filaments to deform elastically by several percent, with energy storage levels of perhaps 10 J/cm^3 for metal nanotubes. Human biceps could be replaced with an 8 mm diameter wire of this material. Such compact "muscle" might allow future robots to outrun and outjump humans.

Sensing

Touch: Current robotic and prosthetic hands receive far less tactile information than the human hand. Recent research has developed a tactile sensor array that mimics the mechanical properties and touch receptors of human fingertips. The sensor array is constructed as a rigid core surrounded by conductive fluid contained by an elastomeric skin. Electrodes are mounted on the surface of the rigid core and are connected to an impedance-measuring device within the core. When the artificial skin touches an object the fluid path around the electrodes is deformed, producing impedance changes that map the forces received from the object. The researchers expect that an important function of such artificial fingertips will be adjusting robotic grip on held objects.

Scientists from several European countries and Israel developed a prosthetic hand in 2009, called Smart Hand, which functions like a real one—allowing patients to write with it, type on a keyboard, play piano and perform other fine movements. The prosthesis has sensors which enable the patient to sense real feeling in its fingertips.

Vision

Computer vision is the science and technology of machines that see. As a scientific discipline, computer vision is concerned with the theory behind artificial systems that extract information from images. The image data can take many forms, such as video sequences and views from cameras.

In most practical computer vision applications, the computers are pre-programmed to solve a particular task, but methods based on learning are now becoming increasingly common.

Computer vision systems rely on image sensors which detect electromagnetic radiation which is typically in the form of either visible light or infra-red light. The sensors are designed using solid-state physics. The process by which light propagates and reflects off surfaces is explained using optics. Sophisticated image sensors even require quantum mechanics to provide a complete understanding of the image formation process.

There is a subfield within computer vision where artificial systems are designed to mimic the processing and behaviour of biological systems, at different levels of complexity. Also, some of the learning-based methods developed within computer vision have their background in biology.

Manipulation

Robots needs to manipulate objects; pick up, modify, destroy, or otherwise have an effect. Thus the “hands” of a robot are often referred to as *end effectors*, while the "arm" is referred to as a *manipulator*. Most robot arms have replaceable effectors, each allowing them to perform some small range of tasks. Some have a fixed manipulator which cannot be replaced, while a few have one very general purpose manipulator, for example a humanoid hand.

For the definitive guide to all forms of robot end-effectors, their design, and usage consult the book "Robot Grippers".

Mechanical Grippers

One of the most common effectors is the gripper. In its simplest manifestation it consists of just two fingers which can open and close to pick up and let go of a range of small objects. Fingers can for example be made of a chain with a metal wire run through it.

Vacuum Grippers

Vacuum grippers are very simple astrictive devices, but can hold very large loads provided the prehension surface is smooth enough to ensure suction.

Pick and place robots for electronic components and for large objects like car windscreens, often use very simple vacuum grippers.

General Purpose Effectors

Some advanced robots are beginning to use fully humanoid hands, like the Shadow Hand, MANUS, and the Schunk hand. These highly dexterous manipulators, with as many as 20 degrees of freedom and hundreds of tactile sensors.

Locomotion

Rolling Robots: For simplicity most mobile robots have four wheels or a number of continuous tracks. Some researchers have tried to create more complex wheeled robots with only one or two wheels. These can have certain advantages such as greater efficiency and reduced parts, as well as allowing a robot to navigate in confined places that a four wheeled robot would not be able to.

Two-Wheeled Balancing Robots

Balancing robots generally use a gyroscope to detect how much a robot is falling and then drive the wheels proportionally in the opposite direction, to counter-balance the fall at hundreds of times per second, based on the dynamics of an inverted pendulum. Many different balancing robots have been designed. While the Segway is not commonly thought of as a robot, it can be thought of as a component of a robot, such as NASA's Robonaut that has been mounted on a Segway.

One-Wheeled Balancing Robots

A one-wheeled balancing robot is an extension of a two-wheeled balancing robot so that it can move in any 2D direction using a round ball as its only wheel. Several one-wheeled balancing robots have been designed recently, such as Carnegie Mellon University's "Ballbot" that is the approximate height and width of a person, and Tohoku Gakuin University's "BallIP". Because of the long, thin shape and ability to maneuver in tight spaces, they have the potential to function better than other robots in environments with people.

Spherical Orb Robots

Several attempts have been made in robots that are completely inside a spherical ball, either by spinning a weight inside the ball, or by rotating the outer shells of the sphere. These have also been referred to as an orb bot or a ball bot

Six-Wheeled Robots

Using six wheels instead of four wheels can give better traction or grip in outdoor terrain such as on rocky dirt or grass.

Tracked Robots

Tank tracks provide even more traction than a six-wheeled robot. Tracked wheels behave as if they were made of hundreds of wheels, therefore are very common for outdoor and military robots, where the robot must drive on very rough terrain. However, they are difficult to use indoors such as on carpets and smooth floors. Examples include NASA's Urban Robot "Urbie".

Walking Applied to Robots

Walking is a difficult and dynamic problem to solve. Several robots have been made which can walk reliably on two legs, however none have yet been made which are as robust as a human. Many other robots have been built that walk on more than two legs, due to these robots being significantly easier to construct. Hybrids too have been proposed in movies such as I, Robot, where they walk on 2 legs and switch to 4 (arms+legs) when going to a sprint. Typically, robots on 2 legs can walk well on flat floors and can occasionally walk up stairs. None can walk over rocky, uneven terrain. Some of the methods which have been tried are:

ZMP Technique

The Zero Moment Point (ZMP) is the algorithm used by robots such as Honda's ASIMO. The robot's onboard computer tries to keep the total inertial forces (the combination of earth's gravity and the acceleration and deceleration of walking), exactly opposed by the floor reaction force (the force of the floor pushing back on the robot's foot). In this way, the two forces cancel out, leaving no moment (force causing the robot to rotate and fall over). However, this is not exactly how a human walks, and the difference is obvious to human observers, some of whom have pointed out that ASIMO walks as if it needs the lavatory. ASIMO's walking algorithm is not static, and some dynamic balancing is used. However, it still requires a smooth surface to walk on.

Hopping

Several robots, built in the 1980s by Marc Raibert at the MIT Leg Laboratory, successfully demonstrated very dynamic walking. Initially, a robot with only one leg, and a very small foot, could stay upright simply by hopping. The movement is the same as that of a person on a pogo stick. As the robot falls to one side, it would jump slightly in that direction, in order to catch itself. Soon, the algorithm was generalised to two and four legs. A bipedal robot was demonstrated running and even performing somersaults. A quadruped was also demonstrated which could trot, run, pace, and bound. For a full list of these robots.

Dynamic Balancing (Controlled Falling)

A more advanced way for a robot to walk is by using a dynamic balancing algorithm, which is potentially more robust than the Zero Moment Point technique, as it constantly monitors the robot's motion, and places the feet in order to maintain stability. This technique was recently demonstrated by Anybots' Dexter Robot, which is so stable, it can even jump. Another example is the TU Delft Flame.

Passive Dynamics

Perhaps the most promising approach utilises passive dynamics where the momentum of swinging limbs is used for greater efficiency. It has been shown that totally unpowered humanoid mechanisms can walk down a gentle slope, using only gravity to propel themselves. Using this technique, a robot need only supply a small amount of motor power to walk along a flat surface or a little more to walk up a hill. This technique promises to make walking robots at least ten times more efficient than ZMP walkers, like ASIMO.

Other Methods of Locomotion

Flying

A modern passenger airliner is essentially a flying robot, with two humans to manage it. The autopilot can control the plane for each stage of the journey, including takeoff, normal flight, and even landing. Other flying robots are uninhabited, and are known as unmanned aerial vehicles (UAVs). They can be smaller and lighter without a human pilot onboard, and fly into dangerous territory for military surveillance missions. Some can even fire on targets under command. UAVs are also being developed which can fire on targets automatically, without the need for a command from a human. Other flying robots include cruise missiles, the Entomopter, and the Epson micro helicopter robot. Robots such as the Air Penguin, Air Ray, and Air Jelly have lighter-than-air bodies, propelled by paddles, and guided by sonar.

Snaking

Several snake robots have been successfully developed. Mimicking the way real snakes move, these robots can navigate very confined spaces, meaning they may one day be used to search for people trapped in collapsed buildings. The Japanese ACM-R5 snake robot can even navigate both on land and in water.

Skating

A small number of skating robots have been developed, one of which is a multi-mode walking and skating device. It has four legs, with unpowered wheels, which can either step or roll. Another robot, Plen, can use a miniature skateboard or rollerskates, and skate across a desktop.

Climbing

Several different approaches have been used to develop robots that have the ability to climb vertical surfaces. One approach mimicks the movements of a human climber on a wall with protrusions; adjusting the center of mass and moving each limb in turn to gain leverage. An example of this is Capuchin, built by Stanford University, California. Another approach uses the specialised toe pad method of wall-climbing geckoes, which can run on smooth surfaces such as vertical glass. Examples of this approach include Wallbot and Stickybot. China's "Technology Daily" November 15, 2008 reported New Concept Aircraft (ZHUHAI) Co., Ltd. Dr. Li Hiu Yeung and his research group have recently successfully developed the bionic gecko robot "Speedy Freelander". According to Dr.

Li introduction, this gecko robot can rapidly climbing up and down in a variety of building walls, ground and vertical wall fissure or walking upside down on the ceiling, it is able to adapt on smooth glass, rough or sticky dust walls as well as the various surface of metallic materials and also can automatically identify obstacles, circumvent the bypass and flexible and realistic movements. Its flexibility and speed are comparable to the natural gecko. A third approach is to mimick the motion of a snake climbing a pole.

Swimming (Like a Fish)

It is calculated that when swimming some fish can achieve a propulsive efficiency greater than 90%. Furthermore, they can accelerate and maneuver far better than any man-made boat or submarine, and produce less noise and water disturbance. Therefore, many researchers studying underwater robots would like to copy this type of locomotion. Notable examples are the Essex University Computer Science Robotic Fish, and the Robot Tuna built by the Institute of Field Robotics, to analyse and mathematically model thunniform motion. The Aqua Penguin, designed and built by Festo of Germany, copies the streamlined shape and propulsion by front "flippers" of penguins. Festo have also built the Aqua Ray and Aqua Jelly, which emulate the locomotion of manta ray, and jellyfish, respectively.

Environmental Interaction and Navigation

Though a significant percentage of robots in commission today are either human controlled, or operate in a static environment, there is an increasing interest in robots that can operate autonomously in a dynamic environment. These robots require some combination of navigation hardware and software in order to traverse their environment. In particular unforeseen events (e.g. people and other obstacles that are not stationary) can cause problems or collisions. Some highly advanced robots as ASIMO, EveR-1, Meinü robot have particularly good robot navigation hardware and software.

Also, self-controlled cars, Ernst Dickmanns' driverless car, and the entries in the DARPA Grand Challenge, are capable of sensing the environment well and subsequently making navigational decisions based on this information. Most of these robots employ a GPS navigation device with waypoints, along with radar, sometimes combined with other sensory data such as LIDAR, video cameras, and inertial guidance systems for better navigation between waypoints.

Human-Robot Interaction

If robots are to work effectively in homes and other non-industrial environments, the way they are instructed to perform their jobs, and especially how they will be told to stop will be of critical importance. The people who interact with them may have little or no training in robotics, and so any interface will need to be extremely intuitive.

Science fiction authors also typically assume that robots will eventually be capable of communicating with humans through speech, gestures, and facial expressions, rather than a command-line interface. Although speech would be the most natural way for the human to communicate, it is unnatural for the robot. It will probably be a long time before robots interact as naturally as the fictional C-3PO.

Speech Recognition

Interpreting the continuous flow of sounds coming from a human, in real time, is a difficult task for a computer, mostly because of the great variability of speech. The same word, spoken by the same person may sound different depending on local acoustics, volume, the previous word, whether or not the speaker has a cold, etc.. It becomes even harder when the speaker has a different accent. Nevertheless, great strides have been made in the field since Davis, Biddulph, and Balashek designed the first "voice input system" which recognised "ten digits spoken by a single user with 100% accuracy" in 1952. Currently, the best systems can recognise continuous, natural speech, up to 160 words per minute, with an accuracy of 95%.

Robotic Voice

Other hurdles exist when allowing the robot to use voice for interacting with humans. For social reasons, synthetic voice proves suboptimal as a communication medium, making it necessary to develop the emotional component of robotic voice through various techniques.

Gestures

One can imagine, in the future, explaining to a robot chef how to make a pastry, or asking directions from a robot police officer. In both of these cases, making hand gestures would aid the verbal descriptions. In the first case, the robot would be recognising gestures made by the human, and perhaps repeating them for confirmation. In the second case, the robot police officer would gesture to indicate "down the road, then turn right". It is likely that gestures will make up a part of the interaction between humans and robots. A great many systems have been developed to recognise human hand gestures.

Facial Expression

Facial expressions can provide rapid feedback on the progress of a dialog between two humans, and soon it may be able to do the same for humans and robots. Robotic faces have been constructed by Hanson Robotics using their elastic polymer called Frubber, allowing a great amount of facial expressions due to the elasticity of the rubber facial coating and imbedded subsurface motors (servos) to produce the facial expressions. The coating and servos are built on a metal skull. A robot should know how to approach a human, judging by their facial expression and body language. Whether the person is happy, frightened, or crazy-looking affects the type of interaction expected of the robot. Likewise, robots like Kismet and the more recent addition, Nexi can produce a range of facial expressions, allowing it to have meaningful social exchanges with humans.

Artificial Emotions

Artificial emotions can also be imbedded and are composed of a sequence of facial expressions and/or gestures. As can be seen from the movie Final Fantasy: The Spirits Within, the programming of these artificial emotions is complex and requires a great amount of human observation. To simplify this programming in the movie, presets were created together with a special software programme. This decreased the amount of time needed to make the film. These presets could possibly be transferred for use in real-life robots.

Personality

Many of the robots of science fiction have a personality, something which may or may not be desirable in the commercial robots of the future. Nevertheless, researchers are trying to create robots which appear to have a personality: i.e. they use sounds, facial expressions, and body language to try to convey an internal state, which may be joy, sadness, or fear. One commercial example is Pleo, a toy robot dinosaur, which can exhibit several apparent emotions.

Control

The mechanical structure of a robot must be controlled to perform tasks. The control of a robot involves three distinct phases - perception, processing, and action (robotic paradigms). Sensors give information about the environment or the robot itself (e.g. the position of its joints or its end effector). This information is then processed to calculate the appropriate signals to the actuators (motors) which move the mechanical.

The processing phase can range in complexity. At a reactive level, it may translate raw sensor information directly into actuator commands. Sensor fusion may first be used to estimate parameters of interest (e.g. the position of the robot's gripper) from noisy sensor data. An immediate task (such as moving the gripper in a certain direction) is inferred from these estimates. Techniques from control theory convert the task into commands that drive the actuators.

At longer time scales or with more sophisticated tasks, the robot may need to build and reason with a "cognitive" model. Cognitive models try to represent the robot, the world, and how they interact. Pattern recognition and computer vision can be used to track objects. Mapping techniques can be used to build maps of the world. Finally, motion planning and other artificial intelligence techniques may be used to figure out how to act. For example, a planner may figure out how to achieve a task without hitting obstacles, falling over, etc.

Autonomy Levels

Control systems may also have varying levels of autonomy.

1. Direct interaction is used for haptic or tele-operated devices, and the human has nearly complete control over the robot's motion.
2. Operator-assist modes have the operator commanding medium-to-high-level tasks, with the robot automatically figuring out how to achieve them.
3. An autonomous robot may go for extended periods of time without human interaction. Higher levels of autonomy do not necessarily require more complex cognitive capabilities. For example, robots in assembly plants are completely autonomous, but operate in a fixed pattern.

Another classification takes into account the interaction between human control and the machine motions.

1. Teleoperation. A human controls each movement, each machine actuator change is specified by the operator.
2. Supervisory. A human specifies general moves or position changes and the machine decides specific movements of its actuators.
3. Task-level autonomy. The operator specifies only the task and the robot manages itself to complete it.
4. Full autonomy. The machine will create and complete all its tasks without human interaction.

Robotics Research

Much of the research in robotics focuses not on specific industrial tasks, but on investigations into new types of robots, alternative ways to think about or design robots, and new ways to manufacture them but other investigations, such as MIT's cyberflora project, are almost wholly academic.

A first particular new innovation in robot design is the opensourcing of robot-projects. To describe the level of advancement of a robot, the term "Generation Robots" can be used. This term is coined by Professor Hans Moravec, Principal Research Scientist at the Carnegie Mellon University Robotics Institute in describing the near future evolution of robot technology. *First generation* robots, Moravec predicted in 1997, should have an intellectual capacity comparable to perhaps a lizard and should become available by 2010. Because the *first generation* robot would be incapable of learning, however, Moravec predicts that the *second generation* robot would be an improvement over the *first* and become available by 2020, with an intelligence maybe comparable to that of a mouse. The *third generation* robot should have an intelligence comparable to that of a monkey. Though *fourth generation* robots, robots with human intelligence, professor Moravec predicts, would become possible, he does not predict this happening before around 2040 or 2050.

The second is Evolutionary Robots. This is a methodology that uses evolutionary computation to help design robots, especially the body form, or motion and behaviour controllers. In a similar way to natural evolution, a large population of robots is allowed to compete in some way, or their ability to perform a task is measured using a fitness function. Those that perform worst are removed from the population, and replaced by a new set, which have new behaviours based on those of the winners. Over time the population improves, and eventually a satisfactory robot may appear. This happens without any direct programming of the robots by the researchers. Researchers use this method both to create better robots, and to explore the nature of evolution. Because the process often requires many generations of robots to be simulated, this technique may be run entirely or mostly in simulation, then tested on real robots once the evolved algorithms are good enough. Currently, there are about 1 million industrial robots toiling around the world, and Japan is the top country having high density of utilising robots in its manufacturing industry.

Dynamics and Kinematics

The study of motion can be divided into kinematics and dynamics. Direct kinematics refers to the calculation of end effector position, orientation, velocity, and acceleration when the corresponding joint values are known. Inverse kinematics refers to the opposite case in which required joint values are calculated for given end effector values, as done in path planning. Some special aspects of kinematics include handling of redundancy (different possibilities of performing the same movement), collision avoidance, and singularity avoidance. Once all relevant positions, velocities, and accelerations have been calculated using kinematics, methods from the field of dynamics are used to study the effect of forces upon these movements. Direct dynamics refers to the calculation of accelerations in the robot once the applied forces are known. Direct dynamics is used in computer simulations of the robot. Inverse dynamics refers to the calculation of the actuator forces necessary to create a prescribed end effector acceleration. This information can be used to improve the control algorithms of a robot.

In each area mentioned above, researchers strive to develop new concepts and strategies, improve existing ones, and improve the interaction between these areas. To do this, criteria for "optimal" performance and ways to optimise design, structure, and control of robots must be developed and implemented.

Education and Training

Robots have become a popular educational tool in some middle and high schools, raising interests in computing among students. First-year computer science courses at several universities now include programming of a robot in addition to traditional software engineering-based coursework.

Career Training

Universities offer Bachelors, Masters, and Doctoral degrees in the field of robotics. Select Private Career Colleges and vocational schools offer robotics training to train individuals towards being job-ready and employable in the emerging robotics industry.

Certification

The Robotics Certification Standards Alliance (RCSA) is an international robotics certification authority that confers various industry- and educational-related robotics certifications.

Employment

Robotics is an essential component in any modern manufacturing environment. As factories increase their use of robots, the number of robotics related jobs grow and have been observed to be on a steady rise.

Effects on Unemployment

Some analysts, such as Martin Ford, argue that robots and other forms of automation will ultimately result in significant unemployment as machines begin to match and exceed the capability of workers to perform most jobs. At present the negative impact is only on menial and repetitive jobs, and there is actually a positive impact on the number of jobs for highly skilled technicians, engineers, and specialists. However, these highly skilled jobs are not sufficient in number to offset the greater decrease in employment among the general population, causing structural unemployment in which overall (net) unemployment rises.

As robotics and artificial intelligence develop further, some worry even many skilled jobs may be threatened. In conventional economic theory this should merely cause an increase in the productivity of the involved industries, resulting in higher demand for other goods, and hence higher labour demand in these sectors, off-setting whatever negatives are caused. Conventional theory describes the past well but may not describe the future due to shifts in the parameter values that shape the context.

Mechatronics

Mechatronics is the combination of Mechanical engineering, Electronic engineering, Computer engineering, Software engineering, Control engineering, and Systems Design engineering in order to design, and manufacture useful products. Mechatronics is a multidisciplinary field of engineering, that is to say it rejects splitting engineering into separate disciplines. Originally, mechatronics just included the combination between mechanics and electronics, hence the word is only a portmanteau of mechanics and electronics. However, as technical systems have become more and more complex the word has been "updated" during recent years to include more technical areas.

French standard NF E 01-010 gives the following definition: "approach aiming at the synergistic integration of mechanics, electronics, control theory, and computer science within product design and manufacturing, in order to improve and/or optimise its functionality".

Description

A mechatronics engineer unites the principles of mechanics, electronics, and computing to generate a simpler, more economical and reliable system. Mechatronics is centred on mechanics, electronics, computing, control engineering, molecular engineering (from nanochemistry and biology), and optical engineering, which, combined, make possible the generation of simpler, more economical, reliable and versatile systems. The portmanteau "mechatronic" was coined by Tetsuro Mori, the senior engineer of the Japanese company Yaskawa in 1969. An industrial robot is a prime example of a mechatronics system; it includes aspects of electronics, mechanical, and computing to do its day-to-day jobs.

Engineering cybernetics deals with the question of control engineering of mechatronic systems. It is used to control or regulate such a system. Through collaboration, the mechatronic modules perform the production goals and inherit flexible and agile manufacturing properties in the production scheme. Modern production equipment consists of mechatronic modules that are integrated according to a control architecture. The most known architectures involve hierarchy, polyarchy, heterarchy, and hybrid. The methods for achieving a technical effect are described by control algorithms, which might or might not utilise formal methods in their design. Hybrid systems important to mechatronics include production systems, synergy drives, planetary exploration rovers, automotive subsystems such as anti-lock braking systems and spin-assist, and every-day equipment such as autofocus cameras, video, hard disks, and CD players.

Course Structure

Mechatronic students take courses from across the various fields listed below:

- Mechanical engineering and materials science subjects
- Electronic engineering subjects
- Computer engineering subjects
- Computer science subjects
- Systems and control engineering subjects
- Optomechanics (optical engineering) subjects
- Robotics subjects.

Application

- Machine vision
- Automation and robotics
- Servo-mechanics
- Sensing and control systems
- Automotive engineering, automotive equipment in the design of subsystems such as anti-lock braking systems
- Computer-machine controls, such as computer driven machines like IE CNC milling machines
- Expert systems
- Industrial goods
- Consumer products
- Mechatronics systems
- Medical mechatronics,medical imaging systems
- Structural dynamic systems
- Transportation and vehicular systems
- Mechatronics as the new language of the automobile
- Diagnostic, reliability, and control system techniques
- Computer aided and integrated manufacturing systems
- Computer-aided design
- Engineering and manufacturing systems
- Packaging
- Microcontrollers / PLC's
- Mobile apps.

Physical Implementations

For most mechatronic systems, the main issue is no more how to implement a control system, but how to implement actuators and what is the energy source. Within the mechatronic field, mainly twoe movement: the piezo-electric actuators and motors, or the electromagnetic actuators and motors. Maybe the most famous mechatronics systems are the well known camera autofocus system or camera anti-shake systems.

Concerning the energy sources, most of the applications use batteries. But a new trend is arriving and is the energy harvesting, allowing transforming into electricity mechanical energy from shock, vibration, or thermal energy from thermal variation, and so on.

Variant of the Field

An emerging variant of this field is biomechatronics, whose purpose is to integrate mechanical parts with a human being, usually in the form of removable gadgets such as an exoskeleton. Such an entity is often identified in science fiction as a cyborg. This is the "real-life" version of cyberware. Another emerging variant is Electronical or electronics design centric ECAD/MCAD co-design. Electronical is where the integration and co-design between the design team and design tools of an electronics centric system and the design team and design tools of that systems physical/mechanical enclosure takes place.

Education

Countries offering education in mechatronics are Australia, Brazil, Canada, Chile, Perú, Colombia, Czechia, Denmark, Egypt France, Germany, Hungary, India, Iran, Ireland, Israel Italy, Iceland, Japan, Jordan, Malaysia, México, Norway, Pakistan, Poland, Singapore, Slovenia, Sweden, Turkey, UK, United States, New Zealand, VietNam,among others.

4

Applied Mechanics

Applied mechanics is a branch of the physical sciences and the practical application of mechanics. Applied mechanics examines the *response of bodies (solids and fluids) or systems of bodies to external forces.* Some examples of *mechanical systems* include the flow of a liquid under pressure, the fracture of a solid from an applied force, or the vibration of an ear in response to sound. A practitioner of the discipline is known as a mechanician.

Applied mechanics, as its name suggests, bridges the gap between physical theory and its application to technology. As such, applied mechanics is used in many fields of engineering, especially mechanical engineering. In this context, it is commonly referred to as engineering mechanics. Much of modern engineering mechanics is based on Isaac Newton's laws of motion while the modern practice of their application can be traced back to Stephen Timoshenko, who is said to be the father of modern engineering mechanics.

Within the theoretical sciences, applied mechanics is useful in formulating new ideas and theories, discovering and interpreting phenomena, and developing experimental and computational tools. In the application of the natural sciences, mechanics was said to be complemented by thermodynamics by physical chemists Gilbert N. Lewis and Merle Randall, the study of heat and more generally energy, and electromechanics, the study of electricity and magnetism.

Applied Mechanics in Practice

As a scientific discipline, applied mechanics derives many of its principles and methods from the Physical sciences (in particular,

Mechanics and Classical Mechanics), from Mathematics and, increasingly, from Computer Science. As such, Applied Mechanics shares similar methods, theories, and topics with Applied Physics, Applied Mathematics, and Computational Science.

As an enabling discipline, applied mechanics has received impetus from the study of natural phenomena such as orbits of planets, circulation of blood, locomotion of animals, crawling of cells, formation of mountains, and propagation of seismic waves. Such studies have resulted in disciplines such as celestial mechanics, biomechanics and geomechanics.

As a practical discipline, applied mechanics has also advanced by participating in major inventions throughout history, such as buildings, ships, automobiles, railways, petroleum refineries, engines, airplanes, nuclear reactors, composite materials, computers, and medical implants. In such connections, the discipline is also known as *Engineering Mechanics*, often practiced within Civil Engineering, Mechanical Engineering, Construction Engineering, Materials Science and Engineering, Aerospace Engineering, Chemical Engineering, Electrical Engineering, Nuclear Engineering, Structural engineering and Bioengineering.

Applied Mechanics in Engineering

Typically, engineering mechanics is used to analyse and predict the acceleration and deformation (both elastic and plastic) of objects under known forces (also called loads) or stresses.

When treated as an area of study within a larger engineering curriculum, engineering mechanics can be subdivided into

- Statics, the study of non-moving bodies under known loads
- Dynamics (or kinetics), the study of how forces affect moving bodies
- Mechanics of materials or strength of materials, the study of how different materials deform under various types of stress
- Deformation mechanics, the study of deformations typically in the elastic range
- Fluid mechanics, the study of how fluids react to forces. Note that fluid mechanics can be further split into fluid statics and fluid dynamics, and is itself a subdiscipline of continuum mechanics. The application of fluid mechanics in engineering is called hydraulics.

- Continuum mechanics is a method of applying mechanics that assumes that all objects are continuous. It is contrasted by discrete mechanics.

Biomechanics

Biomechanics is the application of mechanical principles to biological systems, such as humans, animals, plants, organs, and cells. Perhaps one of the best definitions was provided by Herbert Hatze in 1974: "Biomechanics is the study of the structure and function of biological systems by means of the methods of mechanics". The word biomechanics developed during the early 1970s, describing the application of engineering mechanics to biological and medical systems.

Biomechanics is closely related to engineering, because it often uses traditional engineering sciences to analyse biological systems. Some simple applications of Newtonian mechanics and/or materials sciences can supply correct approximations to the mechanics of many biological systems. Applied mechanics, most notably mechanical engineering disciplines such as continuum mechanics, mechanism analysis, structural analysis, kinematics and dynamics play prominent roles in the study of biomechanics.

Usually biological system are more complex than man-built systems. Numerical methods are hence applied in almost every biomechanical study. Research is done in a iterative process of hypothesis and verification, including several steps of modelling, computer simulation and experimental measurements.

Subfields

Plant Biomechanics: The application of biomechanical principles to plants and plant organs has developed into the subfield of plant biomechanics.

Sport Biomechanics: In sports biomechanics, the laws of mechanics are applied in order to gain a greater understanding of athletic performance and to reduce sport injuries as well. Elements of mechanical engineering (e.g., strain gauges), electrical engineering (e.g., digital filtering), computer science (e.g., numerical methods), gait analysis (e.g., force platforms), and clinical neurophysiology (e.g., surface EMG) are common methods used in sports biomechanics.

Continuum Biomechanics: The mechanical analysis of biomaterials and biofluids is usually carried forth with the concepts of continuum mechanics. This assumption breaks down when the length scales of interest approach the order of the micro structural details of the material. One of the most remarkable characteristic of biomaterials is their hierarchical structure. In other words, the mechanical characteristics of these materials rely on physical phenomena occurring in multiple levels, from the molecular all the way up to the tissue and organ levels.

Biomaterials are classified in two groups, hard and soft tissues. Mechanical deformation of hard tissues (like wood, shell and bone) may be analysed with the theory of linear elasticity. On the other hand, soft tissues (like skin, tendon, muscle and cartilage) usually undergo large deformations and thus their analysis rely on the finite strain theory and computer simulations. The interest in continuum biomechanics is spurred by the need for realism in the development of medical simulation.

Comparative Biomechanics: Comparative biomechanics is the application of biomechanics to non-human organisms, whether used to gain greater insights into humans (as in physical anthropology) or into the functions, ecology and adaptations of the organisms themselves. Common areas of investigation are Animal locomotion and feeding, as these have strong connections to the organism's fitness and impose high mechanical demands. Animal locomotion, has many manifestations, including running, jumping and flying. Locomotion requires energy to overcome friction, drag, inertia, and gravity, though which factor predominates varies with environment.

Comparative biomechanics overlaps strongly with many other fields, including ecology, neurobiology, developmental biology, ethology, and paleontology, to the extent of commonly publishing papers in the journals of these other fields. Comparative biomechanics is often applied in medicine (with regards to common model organisms such as mice and rats) as well as in biomimetics, which looks to nature for solutions to engineering problems.

Biofluid Mechanics: Under most circumstances, blood flow can be modelled by the Navier-Stokes equations. Whole blood can often be assumed to be an incompressible Newtonian fluid. However, this assumption fails when considering flows within arterioles. At this scale, the effects of individual red blood cells becomes significant, and whole blood can no longer be modelled as a continuum. When the diameter of the blood vessel is slightly larger than the diameter of the red blood cell the Fahraeus–Lindquist effect occurs and there is a decrease in wall shear stress. However, as the diameter of the blood vessel decreases further, the red blood cells have to squeeze through the vessel and often can only pass in single file. In this case, the inverse Fahraeus–Lindquist effect occurs and the wall shear stress increases.

Biotribology: The main aspects of tribology are related with friction, wear and lubrication. When the two surfaces come in contact during motion i.e. rub against each other, friction, wear and lubrication effects are very important to analyse in order to determine the performance of the material. Biotribology is a study of friction, wear and lubrication of biological systems especially human joints such as hips and knees. For example, femoral component and tibial component of knee implant rub against each other during daily activity such as walking or stair climbing. If the performance of tibial component needs to be analysed, the principles of biotribology are used to determine the wear performance of the implant and lubrication effects of synovial fluid. In addition, the theory of contact mechanics also becomes very important for wear analysis.

History

Antiquity: Aristotle wrote the first book on biomechanics, *De Motu Animalium*, or On the Movement of Animals. He not only saw animals' bodies as mechanical systems, but pursued questions such as the physiological difference between imagining performing an action and actually doing it. Some simple examples of biomechanics research include the investigation of the forces that act on limbs, the aerodynamics of bird and insect flight, the hydrodynamics of swimming in fish, and locomotion in general across all forms of life, from individual cells to whole organisms. The biomechanics of human beings is a core part of kinesiology.

Renaissance: Probably Leonardo da Vinci could be recognised as the first true biomechanician, because he was the first to study anatomy in the context of mechanics. He analysed muscle forces as acting along lines connecting origins and insertions and studied joint function. He also intended to mimic some animal features in his machines. For example, he studied the flight of birds to find means by which humans could fly. Because horses were the principal source of mechanical power in that time, he studied their muscular systems to design machines that would better benefit from the forces applied by this animal.

Galileo Galilei was interested in the strength of bones and suggested that bones are hollow for this affords maximum strength with minimum weight. He noted that animals' masses increase disproportionately to their size, and their bones must consequently also disproportionately increase in girth, adapting to loadbearing rather than mere size the bending strength of a tubular structure such as a bone is increased relative to its weight. This surely was one of the first grasps of principles of biological optimisation.

In the 16th century, Descartes suggested a philosophic system whereby all living systems, including the human body (but not the soul), are simply machines ruled by the same mechanical laws, an idea that did much to promote and sustain biomechanical study. Giovanni Alfonso Borelli embraced this idea and studied walking, running, jumping, the flight of birds, the swimming of fish, and even the piston action of the heart within a mechanical framework. He could determine the position of the human center of gravity, calculate and measured inspired and expired air volumes, and showed that inspiration is muscle-driven and expiration is due to tissue elasticity. Borelli was the first to understand that the levers of the musculoskeletal system magnify motion rather than force, so that muscles must produce much larger forces than those resisting the motion. Influenced by the work of Galileo, whom he personally knew, he had an intuitive understanding of static equilibrium in various joints of the human body well before Newton published the laws of motion.

Industrial Era: In the 19th century Etienne-Jules Marey used cinematography to scientifically investigate locomotion. He opened the field of modern 'motion analysis' by being the first to correlate ground reaction forces with movement. In Germany, the brothers Ernst Heinrich Weber and Wilhelm Eduard Weber hypothesised a great deal about human gait, but it was Christian Wilhelm Braune who significantly advanced the science using recent advances in engineering mechanics. During the same period, the engineering mechanics of materials began to flourish in France and Germany under the demands of the industrial revolution. This led to the rebirth of bone biomechanics when the railroad engineer Karl Culmann and the anatomist Hermann von Meyer compared the stress patterns in a human femur with those in a similarly shaped crane. Inspired by this finding Julius Wolff proposed the famous Wolff's law of bone remodelling.

Applications

The study of biomechanics ranges from the inner workings of a cell to the movement and development of limbs, to the mechanical properties of soft tissue, and bones. As we develop a greater understanding of the physiological behaviour of living tissues, researchers are able to advance the field of tissue engineering, as well as develop improved treatments for a wide array of pathologies.

Biomechanics is also applied to studying human musculoskeletal systems. Such research utilises force platforms to study human ground reaction forces and infrared videography to capture the trajectories of markers attached to the human body to study human 3D motion. Research also applies electromyography (EMG) system to study the muscle activation. By this, it is feasible to investigate the muscle responses to the external forces as well as perturbations.

Biomechanics is widely used in orthopedic industry to design orthopedic implants for human joints, dental parts, external fixations and other medical purposes. Biotribology is a very important part of it. It is a study of the performance and function of biomaterials used for orthopedic implants. It plays a vital role to improve the design and produce successful biomaterials for medical and clinical purposes.

Statistical Mechanics

Statistical mechanics or statistical thermodynamics is a branch of physics that applies probability theory, which contains mathematical tools for dealing with large populations, to the study of the *thermodynamic* behaviour of systems composed of a *large* number of particles. Statistical mechanics provides a framework for relating the microscopic properties of individual atoms and molecules to the macroscopic bulk properties of materials that can be observed in everyday life, therefore explaining thermodynamics as a result of classical and quantum-mechanical description of statistics and mechanics at the microscopic level.

Statistical mechanics provides a molecular-level interpretation of macroscopic thermodynamic quantities such as work, heat, free energy, and entropy. It enables the thermodynamic properties of bulk materials to be related to the spectroscopic data of individual molecules. This ability to make macroscopic predictions based on microscopic properties is the main advantage of statistical mechanics over classical thermodynamics. Both theories are governed by the second law of thermodynamics through the medium of entropy. However, entropy in thermodynamics can only be known empirically, whereas in statistical mechanics, it is a function of the distribution of the system on its micro-states.

Statistical mechanics was initiated in 1870 with the work of Austrian physicist Ludwig Boltzmann, much of which was collectively published in Boltzmann's 1896 *Lectures on Gas Theory*. Boltzmann's original papers on the statistical interpretation of thermodynamics, the H-theorem, transport theory, thermal equilibrium, the equation of state of gases, and similar subjects, occupy about 2,000 pages in the proceedings of the Vienna Academy and other societies. The term "statistical thermodynamics" was proposed for use by the American thermodynamicist and physical chemist J. Willard Gibbs in 1902. According to Gibbs, the term "statistical", in the context of mechanics, i.e. statistical mechanics, was first used by the Scottish physicist James Clerk Maxwell in 1871. "Probabilistic mechanics" might today seem a more appropriate term, but "statistical mechanics" is firmly entrenched.

Overview

The essential problem in statistical thermodynamics is to calculate the distribution of a given amount of energy E over N identical systems. The goal of statistical thermodynamics is to understand and to interpret the measurable macroscopic properties of materials in terms of the properties of their constituent particles and the interactions between them. This is done by connecting thermodynamic functions to quantum-mechanical equations. Two central quantities in statistical thermodynamics are the Boltzmann factor and the partition function.

Fundamentals

Central topics covered in statistical thermodynamics include:

- Microstates and configurations
- Boltzmann distribution law
- Partition function, Configuration integral or configurational partition function
- Thermodynamic equilibrium - thermal, mechanical, and chemical.
- Internal degrees of freedom - rotation, vibration, electronic excitation, etc.
- Heat capacity – Einstein solids, polyatomic gases, etc.
- Nernst heat theorem
- Fluctuations
- Gibbs paradox
- Degeneracy.

Lastly, and most importantly, the formal definition of entropy of a thermodynamic system from a statistical perspective is called statistical entropy, and is defined as:

$$S = k_B \ln \Omega$$

where

k_B is Boltzmann's constant 1.38066×10^{-23} J K^{-1} and

Ω is the number of microstates corresponding to the observed thermodynamic macrostate.

This equation is valid only if each microstate is equally accessible (each microstate has an equal probability of occurring).

Boltzmann Distribution

Some or all of the formulas presented have missing or incomplete descriptions of their variables, symbols or constants, which may create ambiguity or prevent full interpretation.

History

In 1738, Swiss physicist and mathematician Daniel Bernoulli published *Hydrodynamica* which laid the basis for the kinetic theory of gases. In this work, Bernoulli posited the argument, still used to this day, that gases consist of great numbers of molecules moving in all directions, that their impact on a surface causes the gas pressure that we feel, and that what we experience as heat is simply the kinetic energy of their motion.

In 1859, after reading a paper on the diffusion of molecules by Rudolf Clausius, Scottish physicist James Clerk Maxwell formulated the Maxwell distribution of molecular velocities, which gave the proportion of molecules having a certain velocity in a specific range. This was the first-ever statistical law in physics. Five years later, in 1864, Ludwig Boltzmann, a young student in Vienna, came across Maxwell's paper and was so inspired by it that he spent much of his life developing the subject further.

Hence, the foundations of statistical thermodynamics were laid down in the late 1800s by those such as Maxwell, Boltzmann, Max Planck, Clausius, and Josiah Willard Gibbs who began to apply statistical and quantum atomic theory to ideal gas bodies. Predominantly, however, it was Maxwell and Boltzmann, working independently, who reached similar conclusions as to the statistical nature of gaseous bodies. Yet, one must consider Boltzmann to be the "father" of statistical thermodynamics with his 1875 derivation of the relationship between entropy S and multiplicity Ω, the number of microscopic arrangements (microstates) producing the same macroscopic state (macrostate) for a particular system.

Fundamental Postulate

The fundamental postulate in statistical mechanics (also known as the *equal a priori probability postulate*) is the following:

Given an isolated system in equilibrium, it is found with equal probability in each of its accessible microstates.

This postulate is a fundamental assumption in statistical mechanics - it states that a system in equilibrium does not have any preference for any of its available microstates. Given Ω microstates at a particular energy, the probability of finding the system in a particular microstate is $p = 1/\Omega$.

This postulate is necessary because it allows one to conclude that for a system at equilibrium, the thermodynamic state (macrostate) which could result from the largest number of microstates is also the most probable macrostate of the system.

The postulate is justified in part, for classical systems, by Liouville's theorem (Hamiltonian), which shows that if the distribution of system points through accessible phase space is uniform at some time, it remains so at later times.

Similar justification for a discrete system is provided by the mechanism of detailed balance.

This allows for the definition of the *information function* (in the context of information theory):

$$I = -\sum_i \rho_i \ln \rho_i = \langle -\ln \rho \rangle.$$

When all the probabilities (ρ_i) are equal, I is maximal, and we have minimal information about the system. When our information is maximal (i.e., one rho is equal to one and the rest to zero, such that we know what state the system is in), the function is minimal.

This information function is the same as the *reduced entropic function* in thermodynamics.

Mark Srednicki has argued that the fundamental postulate can be derived assuming only that Berry's conjecture (named after Michael Berry) applies to the system in question. Berry's conjecture is believed to hold only for chaotic systems, and roughly says that the energy eigenstates are distributed as Gaussian random variables. Since all realistic systems with more than a handful of degrees of freedom are expected to be chaotic, this puts the fundamental postulate on firm footing. Berry's conjecture has also be shown to be equivalent to an information theoretic *principle of least bias.*

Statistical Ensembles

The modern formulation of statistical mechanics is based on the description of the physical system by an ensemble that represents all possible configurations of the system and the probability of realising each configuration.

Each ensemble is associated with a partition function that, with mathematical manipulation, can be used to extract values of thermodynamic properties of the system. According to the relationship of the system to the rest of the universe, one of three general types of ensembles may apply, in order of increasing complexity:

- Microcanonical ensemble: describes a completely isolated system, having constant energy, as it does not exchange energy or mass with the rest of the universe.
- Canonical community: describes a system in thermal equilibrium with its environment. It may only exchange energy in the form of heat with the outside.
- Grand-canonical: used in open systems which exchange energy and mass with the outside.

Microcanonical Ensemble

In microcanonical ensemble N, V and E are fixed. Since the second law of thermodynamics applies to isolated systems, the first case investigated will correspond to this case. The *Microcanonical ensemble* describes an isolated system.

The entropy of such a system can only increase, so that the maximum of its entropy corresponds to an equilibrium state for the system.

Because an isolated system keeps a constant energy, the total energy of the system does not fluctuate. Thus, the system can access only those of its micro-states that correspond to a given value E of the energy. The internal energy of the system is then strictly equal to its energy.

Let us call $\Omega(E)$ the number of micro-states corresponding to this value of the system's energy. The macroscopic state of maximal entropy for the system is the one in which all micro-states are equally likely to occur, with probability $1 / \Omega(E)$, during the system's fluctuations.

$$S = -k_B \sum_{i=1}^{\Omega(E)} \left\{ \frac{1}{\Omega(E)} \ln \frac{1}{\Omega(E)} \right\} = k_B \ln\left(\Omega(E)\right)$$

where

S is the system entropy, and

k_B is Boltzmann's constant.

Canonical Ensemble

In canonical ensemble N, V and T are fixed. Invoking the concept of the canonical ensemble, it is possible to derive the probability P_i that a macroscopic system in thermal equilibrium with its environment, will be in a given microstate with energy E_i according to the Boltzmann distribution:

$$P_i = \frac{e^{-\beta E_i}}{\sum_j^{j_{\max}} e^{-\beta E_j}}$$

where

$$\beta = \frac{1}{k_B T},$$

The temperature T arises from the fact that the system is in thermal equilibrium with its environment. The probabilities of the various microstates must add to one, and the normalisation factor in the denominator is the canonical partition function:

$$Z = \sum_i^{i_{max}} e^{-\beta E_i}$$

where E_i is the energy of the ith microstate of the system. The partition function is a measure of the number of states accessible to the system at a given temperature. Canonical ensemble contains a derivation of Boltzmann's factor and the form of the partition function from first principles.

To sum up, the probability of finding a system at temperature T in a particular state with energy E_i is

$$P_i = \frac{e^{-\beta E_i}}{Z}.$$

Thus the partition function looks like the weight factor for the ensemble.

Thermodynamic Connection

The partition function can be used to find the expected (average) value of any microscopic property of the system, which can then be related to macroscopic variables. For instance, the expected value of the microscopic energy E is *interpreted* as the microscopic definition of the thermodynamic variable internal energy U, and can be obtained by taking the derivative of the partition function with respect to the temperature. Indeed,

$$\langle E \rangle = \frac{\sum_i E_i e^{-\beta E_i}}{Z} = -\frac{1}{Z}\frac{dZ}{d\beta}$$

implies, together with the interpretation of <E> as U, the following microscopic definition of internal energy:

$$U := -\frac{d \ln Z}{d\beta}.$$

The entropy can be calculated by

$$\frac{S}{k} = -\sum_i p_i \ln p_i = \sum_i \frac{e^{-\beta E_i}}{Z}(\beta E_i + \ln Z) = \ln Z + \beta U$$

which implies that

$$-\frac{\ln(Z)}{\beta} = U - TS = F$$

is the free energy of the system or in other words,

$Z = e^{-\beta F}$.

Having microscopic expressions for the basic thermodynamic potentials U (internal energy), S (entropy) and F (free energy) is sufficient to derive expressions for other thermodynamic quantities. The basic strategy is as follows. There may be an intensive or extensive quantity that enters explicitly in the expression for the microscopic energy E_i, for instance magnetic field (intensive) or volume (extensive). Then, the conjugate thermodynamic variables are derivatives of the internal energy. The macroscopic magnetisation (extensive) is the derivative of U with respect to the (intensive) magnetic field, and the pressure (intensive) is the derivative of U with respect to volume (extensive).

The treatment in this section assumes no exchange of matter (i.e. fixed mass and fixed particle numbers). However, the volume of the system is variable which means the density is also variable.

This probability can be used to find the average value, which corresponds to the macroscopic value, of any property, J, that depends on the energetic state of the system by using the formula:

$$\langle J\rangle = \sum_i p_i J_i = \sum_i J_i \frac{e^{-\beta E_i}}{Z}$$

where <J> is the average value of property J. This equation can be applied to the internal energy, U:

$$U = \sum_i E_i \frac{e^{-\beta E_i}}{Z}.$$

Subsequently, these equations can be combined with known thermodynamic relationships between U and V to arrive at an expression for pressure in terms of only temperature, volume and the partition function. Similar relationships in terms of the partition function can be derived for other thermodynamic properties as shown in the following table.

It is often useful to consider the energy of a given molecule to be distributed among a number of modes. For example, translational energy refers to that portion of energy associated with the motion of the center of mass of the molecule. Configurational energy refers to that portion of energy associated with the various attractive and repulsive forces between molecules in a system. The other modes are all considered to be internal to each molecule. They include rotational, vibrational, electronic and nuclear modes. If we assume that each mode is independent (a questionable assumption) the total energy can be expressed as the sum of each of the components:

$$E = E_t + E_c + E_n + E_e + E_r + E_v,$$

where the subscripts t, c, n, e, r, and v correspond to translational, configurational, nuclear, electronic, rotational and vibrational modes, respectively. The relationship in this equation can be substituted into the very first equation to give:

$$Z = \sum_i e^{-\beta(E_{ti}+E_{ci}+E_{ni}+E_{ei}+E_{ri}+E_{vi})}$$

$$= \sum_i e^{-\beta E_{ti}} e^{-\beta E_{ci}} e^{-\beta E_{ni}} e^{-\beta E_{ei}} e^{-\beta E_{ri}} e^{-\beta E_{vi}}.$$

If we can assume all these modes are completely uncoupled and uncorrelated, so all these factors are in a probability sense completely independent, then

$$Z = Z_t Z_c Z_n Z_e Z_r Z_v.$$

Thus a partition function can be defined for each mode. Simple expressions have been derived relating each of the various modes to various measurable molecular properties, such as the characteristic rotational or vibrational frequencies.

Grand Canonical Ensemble

In grand canonical ensemble *V*, *T* and chemical potential are fixed. If the system under study is an open system (in which matter can be exchanged), *but* particle number is not conserved, we would have to introduce chemical potentials, μ_j, j = 1,...,n and replace the canonical partition function with the grand canonical partition function:

$$\Xi(V,T,\mu) = \sum_i \exp\left(\beta\left[\sum_{j=1}^{n}\mu_j N_{ij} - E_i\right]\right)$$

where N_{ij} is the number of jth species particles in the ith configuration. Sometimes, we also have other variables to add to the partition function, one corresponding to each conserved quantity. Most of them, however, can be safely interpreted as chemical potentials. In most condensed matter systems, things are non-relativistic and mass is conserved. However, most condensed matter systems of interest also conserve particle number approximately (metastably) and the mass (non-relativistically) is none other than the sum of the number of each type of particle times its mass. Mass is inversely related to density, which is the conjugate variable to pressure. We will ignore this complication and pretend chemical potentials don't matter.

Equivalence between Descriptions at the Thermodynamic Limit

All of the above descriptions differ in the way they allow the given system to fluctuate between its configurations.

In the micro-canonical ensemble, the system exchanges no energy with the outside world, and is therefore not subject to energy fluctuations; in the canonical ensemble, the system is free to exchange energy with the outside in the form of heat.

In the thermodynamic limit, which is the limit of large systems, fluctuations become negligible, so that all these descriptions converge to the same description. In other words, the macroscopic behaviour of a system does not depend on the particular ensemble used for its description.

Given these considerations, the best ensemble to choose for the calculation of the properties of a macroscopic system is that ensemble which allows the result to be derived most easily.

Random Walks

The study of long chain polymers has been a source of problems within the realms of statistical mechanics since about the 1950s. One of the reasons however that scientists were interested in their study is that the equations governing the behaviour of a polymer chain were independent of the chain chemistry. What is more, the governing equation turns out to be a random walk, or diffusive walk, in space. Indeed, the Schrödinger equation is itself a diffusion equation in imaginary time, $t' = it$.

Random Walks in Time

The first example of a random walk is one in space, whereby a particle undergoes a random motion due to external forces in its surrounding medium. A typical example would be a pollen grain in a beaker of water. If one could somehow "dye" the path the pollen grain has taken, the path observed is defined as a random walk.

Consider a toy problem, of a train moving along a 1D track in the x-direction. Suppose that the train moves either a distance of + or - a fixed distance b, depending on whether a coin lands heads or tails when flipped. Lets start by considering the statistics of the steps the toy train takes (where S_i is the ith step taken):

$\langle S_i \rangle = 0$; due to *a priori* equal probabilities

$$\langle S_i S_j \rangle = b^2 \delta_{ij}.$$

The second quantity is known as the correlation function. The delta is the kronecker delta which tells us that if the indices i and j are different, then the result is 0, but if $i = j$ then the kronecker delta is 1, so the correlation function returns a value of b^2. This makes sense, because if $i = j$ then we are considering the same step. Rather trivially then it can be shown that the average displacement of the train on the x-axis is 0;

$$x = \sum_{i=1}^{N} S_i$$

$$\langle x \rangle = \langle \sum_{i=1}^{N} S_i \rangle$$

$$\langle x \rangle = \sum_{i=1}^{N} \langle S_i \rangle.$$

As stated $\langle S_i \rangle$ is 0, so the sum of 0 is still 0. It can also be shown, using the same method demonstrated above, to calculate the root mean square value of problem. The result of this calculation is given below

$$x_{rms} = \sqrt{\langle x^2 \rangle} = b\sqrt{N}.$$

From the diffusion equation it can be shown that the distance a diffusing particle moves in a media is proportional to the root of the time the system has been diffusing for, where the proportionality constant is the root of the diffusion constant.

The above relation, although cosmetically different reveals similar physics, where N is simply the number of steps moved (is loosely connected with time) and b is the characteristic step length. As a consequence we can consider diffusion as a random walk process.

A Hookian Spring!

This result is known as the *entropic spring result* and amounts to saying that upon stretching a polymer chain you are doing work on the system to drag it away from its (preferred) equilibrium state. An example of this is a common elastic band, composed of long chain (rubber) polymers. By stretching the elastic band you are doing work on the system and the band behaves like a conventional spring. What is particularly astonishing about this result however, is that the work done in stretching the polymer chain can be related entirely to the change in entropy of the system as a result of the stretching.

Classical Thermodynamics Us. Statistical Thermodynamics

As an example, from a classical thermodynamics point of view one might ask what is it about a thermodynamic system of gas molecules, such as ammonia NH_3, that determines the free energy characteristic of that compound? Classical thermodynamics does not provide the answer. If, for example, we were given spectroscopic data, of this body of gas molecules, such as bond length, bond angle, bond rotation, and flexibility of the bonds in NH_3 we should see that the free energy could not be other than it is. To prove this true, we need to bridge the gap between the microscopic realm of atoms and molecules and the macroscopic realm of classical thermodynamics. From physics, statistical mechanics provides such a bridge by teaching us how to conceive of a thermodynamic *system* as an assembly of *units*. More specifically, it demonstrates how the thermodynamic parameters of a system, such as temperature and pressure, are interpretable in terms of the parameters descriptive of such constituent atoms and molecules.

In a bounded system, the crucial characteristic of these microscopic units is that their energies are quantised. That is, where the energies accessible to a macroscopic system form a virtual continuum of possibilities, the energies open to any of its submicroscopic components are limited to a discontinuous set of alternatives associated with integral values of some quantum number.

Particle Physics

Particle physics is a branch of physics that studies the existence and interactions of particles that are the constituents of what is usually referred to as *matter* or *radiation*. In current understanding, particles are excitations of quantum fields and interact following their dynamics. Most of the interest in this area is in *fundamental fields*, each of which cannot be described as a bound state of other fields. The current set of fundamental fields and their dynamics are summarised in a theory called the Standard Model, therefore particle physics is largely the study of the Standard Model's particle content and its possible extensions.

Subatomic Particles

Modern particle physics research is focused on subatomic particles, including atomic constituents such as electrons, protons, and neutrons (protons and neutrons are composite particles called baryons, made of quarks), particles produced by radioactive and scattering processes, such as photons, neutrinos, and muons, as well as a wide range of exotic particles.

To be specific, the term *particle* is a misnomer from classical physics because the dynamics of particle physics are governed by quantum mechanics. As such, they exhibit wave-particle duality, displaying particle-like behaviour under certain experimental conditions and wave-like behaviour in others. In more technical terms, they are described by quantum state vectors in a Hilbert space, which is also treated in quantum field theory. Following the convention of particle physicists, *elementary particles* refer to objects such as electrons and photons as it is well known that these types of particles display wave-like properties as well.

All particles and their interactions observed to date can be described almost entirely by a quantum field theory called the Standard Model. The Standard Model has 17 species of elementary particles: 12 fermions or 24 if distinguishing anti-particles, 4 vector bosons (5 with anti-particles), and 1 scalar boson. These elementary particles can combine to form composite particles, accounting for the hundreds of other species of particles discovered since the 1960s. The Standard Model has been found to agree with almost all the experimental tests conducted to date. However, most particle physicists believe that it is an incomplete description of nature, and that a more fundamental theory awaits discovery. In recent years, measurements of neutrino mass have provided the first experimental deviations from the Standard Model.

Particle physics has impacted the philosophy of science greatly. Some particle physicists adhere to reductionism, a point of view that has been criticised and defended by philosophers and scientists. Other physicists may defend the philosophy of holism, which has commonly been viewed to be reductionism's opposite.

History

The idea that all matter is composed of elementary particles dates to at least the 6th century BC. The philosophical doctrine of atomism and the nature of elementary particles were studied by ancient Greek philosophers such as Leucippus, Democritus and Epicurus; ancient Indian philosophers such as Kanada, Dignâga and Dharmakirti; medieval scientists such as Alhazen, Avicenna, and Algazel; and early modern European physicists such as Pierre Gassendi, Robert Boyle, and Isaac Newton. The particle theory of light was also proposed by Alhazen, Avicenna, Gassendi, and Newton. These early ideas were founded in abstract, philosophical reasoning rather than experimentation and empirical observation.

In the 19th century, John Dalton, through his work on stoichiometry, concluded that each element of nature was composed of a single, unique type of particle. Dalton and his contemporaries believed these were the fundamental particles of nature and thus named them atoms, after the Greek word *atomos*, meaning "indivisible". However, near the end of the century, physicists discovered that atoms are not, in fact, the fundamental particles of nature, but conglomerates of even smaller particles. The early 20th-century explorations of nuclear physics and quantum physics culminated in proofs of nuclear fission in 1939 by Lise Meitner (based on experiments by Otto Hahn), and nuclear fusion by Hans Bethe in the same year. These discoveries gave rise to an active industry of generating one atom from another, even rendering possible (although not profitable) the transmutation of lead into gold (alchemy). They also led to the development of nuclear weapons. Throughout the 1950s and 1960s, a bewildering variety of particles were found in scattering experiments. This was referred to as the "particle zoo". This term was deprecated after the formulation of the Standard Model during the 1970s in which the large number of particles was explained as combinations of a (relatively) small number of fundamental particles.

The Standard Model

The very current state of the classification of elementary particles is the Standard Model. It describes the strong, weak, and electromagnetic fundamental interactions, using mediating gauge bosons. The species of gauge bosons are the gluons, W", W+ and Z bosons, and the photons. The model also contains 24 fundamental particles, which are the constituents of matter. Finally, it predicts the existence of a type of boson known as the Higgs boson, which is yet to be discovered.

Experimental Laboratories

In particle physics, the major international laboratories are:

- Brookhaven National Laboratory (Long Island, United States). Its main facility is the Relativistic Heavy Ion Collider (RHIC), which collides heavy ions such as gold ions and polarised protons. It is the world's first heavy ion collider, and the world's only polarised proton collider.
- Budker Institute of Nuclear Physics (Novosibirsk, Russia)
- CERN, (Franco-Swiss border, near Geneva). Its main project is now the Large Hadron Collider (LHC), which had its first beam circulation on 10 September 2008, and is now the world's most energetic collider of protons. It will also be the most energetic collider of heavy ions when it begins colliding lead ions in 2010. Earlier facilities include the Large Electron–Positron Collider (LEP), which was stopped in 2001 and then dismantled to give way for LHC; and the Super Proton Synchrotron, which is being reused as a pre-accelerator for LHC.
- DESY (Hamburg, Germany). Its main facility is the Hadron Elektron Ring Anlage (HERA), which collides electrons and positrons with protons.
- Fermilab, (Batavia, United States). Its main facility is the Tevatron, which collides protons and antiprotons and was the highest-energy particle collider in the world until the Large Hadron Collider surpassed it on 29 November 2009.
- KEK, (Tsukuba, Japan). It is the home of a number of experiments such as the K2K experiment, a neutrino oscillation experiment and Belle, an experiment measuring the CP violation of B mesons.

Many other particle accelerators exist.

The techniques required to do modern experimental particle physics are quite varied and complex, constituting a sub-specialty nearly completely distinct from the theoretical side of the field.

Theory

Theoretical particle physics attempts to develop the models, theoretical framework, and mathematical tools to understand current experiments and make predictions for future experiments. There are several major interrelated efforts in theoretical particle physics today. One important branch attempts to better understand the Standard Model and its tests. By extracting the parameters of the Standard Model from experiments with less uncertainty, this work probes the limits of the Standard Model and therefore expands our understanding of nature's building blocks. These efforts are made challenging by the difficulty of calculating quantities in quantum chromodynamics. Some theorists working in this area refer to themselves as phenomenologists and may use the tools of quantum field theory and effective field theory. Others make use of lattice field theory and call themselves *lattice theorists.*

Another major effort is in model building where model builders develop ideas for what physics may lie beyond the Standard Model (at higher energies or smaller distances). This work is often motivated by the hierarchy problem and is constrained by existing experimental data. It may involve work on supersymmetry, alternatives to the Higgs mechanism, extra spatial dimensions (such as the Randall-Sundrum models), Preon theory, combinations of these, or other ideas.

A third major effort in theoretical particle physics is string theory. *String theorists* attempt to construct a unified description of quantum mechanics and general relativity by building a theory based on small strings, and branes rather than particles. If the theory is successful, it may be considered a "Theory of Everything".

There are also other areas of work in theoretical particle physics ranging from particle cosmology to loop quantum gravity.

This division of efforts in particle physics is reflected in the names of categories on the arXiv, a preprint archive: hep-th (theory), hep-ph (phenomenology), hep-ex (experiments), hep-lat (lattice gauge theory).

The Future

The overarching goal, which is pursued in several distinct ways, is to find and understand what physics may lie beyond the standard model. There are several powerful experimental reasons to expect new physics, including dark matter and neutrino mass. There are also theoretical hints that this new physics should be found at accessible energy scales. Furthermore, there may be unexpected and unpredicted surprises that will give us the most opportunity to learn about nature.

Much of the efforts to find this new physics are focused on new collider experiments. The Large Hadron Collider (LHC) was completed in 2008 to help continue the search for the Higgs boson, supersymmetric particles, and other new physics. An intermediate goal is the construction of the International Linear Collider (ILC), which will complement the LHC by allowing more precise measurements of the properties of newly found particles. In August 2004, a decision for the technology of the ILC was taken but the site has still to be agreed upon.

In addition, there are important non-collider experiments that also attempt to find and understand physics beyond the Standard Model. One important non-collider effort is the determination of the neutrino masses, since these masses may arise from neutrinos mixing with very heavy particles. In addition, cosmological observations provide many useful constraints on the dark matter, although it may be impossible to determine the exact nature of the dark matter without the colliders. Finally, lower bounds on the very long lifetime of the proton put constraints on Grand Unified Theories at energy scales much higher than collider experiments will be able to probe any time soon.

Nuclear Physics

Nuclear physics is the field of physics that studies the building blocks and interactions of atomic nuclei. The most commonly known applications of nuclear physics are nuclear power generation and nuclear weapons technology, but the research has provided application in many fields, including those in nuclear medicine and magnetic resonance imaging, ion implantation in materials engineering, and radiocarbon dating in geology and archaeology.

The field of particle physics evolved out of nuclear physics and is typically taught in close association with nuclear physics.

History

The history of nuclear physics as a discipline distinct from atomic physics starts with the discovery of radioactivity by Henri Becquerel in 1896, while investigating phosphorescence in uranium salts. The discovery of the electron by J. J. Thomson a year later was an indication that the atom had internal structure. At the turn of the 20th century the accepted model of the atom was J. J. Thomson's *plum pudding* model in which the atom was a large positively charged ball with small negatively charged electrons embedded inside of it. By the turn of the century physicists had also discovered three types of radiation emanating from atoms, which they named alpha, beta, and gamma radiation. Experiments in 1911 by Lise Meitner and Otto Hahn, and by James Chadwick in 1914 discovered that the beta decay spectrum was continuous rather than discrete. That is, electrons were ejected from the atom with a range of energies, rather than the discrete amounts of energies that were observed in gamma and alpha decays. This was a problem for nuclear physics at the time, because it indicated that energy was not conserved in these decays.

In 1905, Albert Einstein formulated the idea of mass–energy equivalence. While the work on radioactivity by Becquerel, Pierre and Marie Curie predates this, an explanation of the source of the energy of radioactivity would have to wait for the discovery that the nucleus itself was composed of smaller constituents, the nucleons.

Rutherford's Team Discovers the Nucleus

In 1907 Ernest Rutherford published "Radiation of the á Particle from Radium in passing through Matter." Geiger expanded on this work in a communication to the Royal Society with experiments he and Rutherford had done passing á particles through air, aluminium foil and gold leaf. More work was published in 1909 by Geiger and Marsden and further greatly expanded work was published in 1910 by Geiger, In 1911-2 Rutherford went before the Royal Society to explain the experiments and propound the new theory of the atomic nucleus as we now understand it.

The key experiment behind this announcement happened in 1910 at the University of Manchester, as Ernest Rutherford's team performed a remarkable experiment in which Hans Geiger and Ernest Marsden under his supervision fired alpha particles (helium nuclei) at a thin film of gold foil. The plum pudding model predicted that the alpha particles should come out of the foil with their trajectories being at most slightly bent. Rutherford had the idea to instruct his team to look for something that shocked him to actually observe: a few particles were scattered through large angles, even completely backwards, in some cases. He likened it to firing a bullet at tissue paper and having it bounce off. The discovery, beginning with Rutherford's analysis of the data in 1911, eventually led to the Rutherford model of the atom, in which the atom has a very small, very dense nucleus containing most of its mass, and consisting of heavy positively charged particles with embedded electrons in order to balance out the charge (since the neutron was unknown). As an example, in this model (which is not the modern one) nitrogen-14 consisted of a nucleus with 14 protons and 7 electrons (21 total particles), and the nucleus was surrounded by 7 more orbiting electrons.

The Rutherford model worked quite well until studies of nuclear spin were carried out by Franco Rasetti at the California Institute of Technology in 1929. By 1925 it was known that protons and electrons had a spin of 1/2, and in the Rutherford model of nitrogen-14, 20 of the total 21 nuclear particles should have paired up to cancel each other's spin, and the final odd particle should have left the nucleus with a net spin of 1/2. Rasetti discovered, however, that nitrogen-14 has a spin of 1.

James Chadwick Discovers the Neutron

In 1932 Chadwick realised that radiation that had been observed by Walther Bothe, Herbert L. Becker, Irène and Frédéric Joliot-Curie was actually due to a neutral particle of about the same mass as the proton, that he called the neutron (following a suggestion about the need for such a particle, by Rutherford). In the same year Dmitri Ivanenko suggested that neutrons were in fact spin 1/2 particles and that the nucleus contained neutrons to explain the mass not due to protons, and that there were no electrons in the nucleus—only protons and neutrons. The neutron spin immediately solved the problem of the spin of nitrogen-14, as the one unpaired proton and one unpaired neutron in this model, each contribute a spin of 1/2 in the same direction, for a final total spin of 1.

With the discovery of the neutron, scientists at last could calculate what fraction of binding energy each nucleus had, from comparing the nuclear mass with that of the protons and neutrons which composed it. Differences between nuclear masses were calculated in this way and—when nuclear reactions were measured—were found to agree with Einstein's calculation of the equivalence of mass and energy to high accuracy (within 1 percent as of in 1934).

Proca's Equations of the Massive Vector Boson Field

Alexandru Proca was the first to develop and report the massive vector boson field equations and a theory of the mesonic field of nuclear forces. Proca's equations were known to Wolfgang Pauli who mentioned the equations in his Nobel address, and they were also known to Yukawa, Wentzel,Taketani, Sakata,Kemmer,Heitler and Fröhlich who appreciated the content of Proca's equations for developing a theory of the atomic nuclei in Nuclear Physics.

Yukawa's Meson Postulated to Bind Nuclei

In 1935 Hideki Yukawa proposed the first significant theory of the strong force to explain how the nucleus holds together. In the Yukawa interaction a virtual particle, later called a meson, mediated a force between all nucleons, including protons and neutrons. This force explained why nuclei did not disintegrate under the influence of proton repulsion, and it also gave an explanation of why the attractive strong force had a more limited range than the electromagnetic repulsion between protons. Later, the discovery of the pi meson showed it to have the properties of Yukawa's particle.

With Yukawa's papers, the modern model of the atom was complete. The center of the atom contains a tight ball of neutrons and protons, which is held together by the strong nuclear force, unless it is too large. Unstable nuclei may undergo alpha decay, in which they emit an energetic helium nucleus, or beta decay, in which they eject an electron (or positron). After one of these decays the resultant nucleus may be left in an excited state, and in this case it decays to its ground state by emitting high energy photons (gamma decay).

The study of the strong and weak nuclear forces (the latter explained by Enrico Fermi via Fermi's interaction in 1934) led physicists to collide nuclei and electrons at ever higher energies. This research became the science of particle physics, the crown jewel of which is the standard model of particle physics which describes the strong, weak, and electromagnetic forces.

Modern Nuclear Physics

A heavy nucleus can contain hundreds of nucleons which means that with some approximation it can be treated as a classical system, rather than a quantum-mechanical one. In the resulting liquid-drop model, the nucleus has an energy which arises partly from surface tension and partly from electrical repulsion of the protons. The liquid-drop model is able to reproduce many features of nuclei, including the general trend of binding energy with respect to mass number, as well as the phenomenon of nuclear fission.

Superimposed on this classical picture, however, are quantum-mechanical effects, which can be described using the nuclear shell model, developed in large part by Maria Goeppert-Mayer. Nuclei with certain numbers of neutrons and protons (the magic numbers 2, 8, 20, 28, 50, 82, 126, ...) are particularly stable, because their shells are filled.

Other more complicated models for the nucleus have also been proposed, such as the interacting boson model, in which pairs of neutrons and protons interact as bosons, analogously to Cooper pairs of electrons.

Much of current research in nuclear physics relates to the study of nuclei under extreme conditions such as high spin and excitation energy. Nuclei may also have extreme shapes (similar to that of Rugby balls) or extreme neutron-to-proton ratios. Experimenters can create such nuclei using artificially induced fusion or nucleon transfer reactions, employing ion beams from an accelerator. Beams with even higher energies can be used to create nuclei at very high temperatures, and there are signs that these experiments have produced a phase transition from normal nuclear matter to a new state, the quark-gluon plasma, in which the quarks mingle with one another, rather than being segregated in triplets as they are in neutrons and protons.

Nuclear Decay

Eighty elements have at least one stable isotope never observed to decay, amounting to a total of about 255 stable isotopes. However, thousands of isotopes have been characterised that are unstable. These radioisotopes decay over time scales ranging from fractions of a second to weeks, years, billions of years, or even trillions of years.

The stability of a nucleus is highest when it falls into a certain range or balance of composition of neutrons and protons; too few or too many neutrons may cause it to decay. For example, in beta decay a nitrogen-16 atom (7 protons, 9 neutrons) is converted to an oxygen-16 atom (8 protons, 8 neutrons) within a few seconds of being created. In this decay a neutron in the nitrogen nucleus is converted into a proton and an electron and an anti-neutrino by the weak nuclear force. The element is transmuted to another element in by acquiring the created proton.

In alpha decay the radioactive element decays by emitting a helium nucleus (2 protons and 2 neutrons), giving another element, plus helium-4. In many cases this process continues through several steps of this kind, including other types of decays, until a stable element is formed.

In gamma decay, a nucleus decays from an excited state into a lower energy state, by emitting a gamma ray. The element is not changed to another element in the process (no nuclear transmutation is involved).

Other more exotic decays are possible. For example, in internal conversion decay, the energy from an excited nucleus may be used to eject one of the inner orbital electrons from the atom, in a process which produces high speed electrons, but is not beta decay, and (unlike beta decay) does not transmute one element to another.

Nuclear Fusion

In nuclear fusion, two low mass nuclei come into very close contact with each other, so that the strong force fuses them. It requires a large amount of energy to overcome the repulsion between the nuclei for the strong or nuclear forces to produce this effect, therefore nuclear fusion can only take place at very high temperatures or high pressures. Once the process succeeds, a very large amount of energy is released and the combined nucleus assumes a lower energy level. The binding energy per nucleon increases with mass number up until nickel-62. Stars like the Sun are powered by the fusion of four protons into a helium nucleus, two positrons, and two neutrinos. The uncontrolled fusion of hydrogen into helium is known as thermonuclear runaway. A frontier in current research at various institutions, for example the Joint European Torus (JET) and ITER, is the development of an economically viable method of using energy from a *controlled* fusion reaction.

Nuclear Fission

Nuclear fission is the reverse process of fusion. For nuclei heavier than nickel-62 the binding energy per nucleon decreases with the mass number. It is therefore possible for energy to be released if a heavy nucleus breaks apart into two lighter ones.

The process of alpha decay is in essence a special type of spontaneous nuclear fission. This process produces a highly asymmetrical fission because the four particles which make up the alpha particle are especially tightly bound to each other, making production of this nucleus in fission particularly likely.

For certain of the heaviest nuclei which produce neutrons on fission, and which also easily absorb neutrons to initiate fission, a self-igniting type of neutron-initiated fission can be obtained, in a so-called chain reaction. Chain reactions were known in chemistry before physics, and in fact many familiar processes like fires and chemical explosions are chemical chain reactions. The fission or "nuclear" chain-reaction, using fission-produced neutrons, is the source of energy for nuclear power plants and fission type nuclear bombs, such as those detonated by the United States in Hiroshima and Nagasaki, Japan, at the end of World War II. Heavy nuclei such as uranium and thorium may undergo spontaneous fission, but they are much more likely to undergo decay by alpha decay.

For a neutron-initiated chain-reaction to occur, there must be a critical mass of the element present in a certain space under certain conditions (these conditions slow and conserve neutrons for the reactions). In two regions of Oklo, Gabon, Africa, natural nuclear fission reactors were active over 1.5 billion years ago. Measurements of natural neutrino emission have demonstrated that around half of the heat emanating from the Earth's core results from radioactive decay. However, it is not known if any of this results from fission chain-reactions.

Production of Heavy Elements

According to the theory, as the Universe cooled after the big bang it eventually became possible for common subatomic particles as we know them (neutrons, protons and electrons) to exist. The most common particles created in the big bang which are still easily observable to us today were protons and electrons (in equal numbers). The protons would eventually form hydrogen atoms. Almost all the neutrons created in the Big Bang were absorbed into helium-4 in the first three minutes after the Big Bang, and this helium accounts for most of the helium in the universe today.

Some fraction of elements beyond helium were created in the Big Bang, as the protons and neutrons collided with each other (lithium, beryllium, and perhaps some boron), but all of the "heavier elements" (heavier than carbon, element number 6) that we see today, were created inside of stars during a series of fusion stages, such as the proton-proton chain, the CNO cycle and the triple-alpha process. Progressively heavier elements are created during the evolution of a star.

Since the binding energy per nucleon peaks around iron, energy is only released in fusion processes occurring below this point. Since the creation of heavier nuclei by fusion costs energy, nature resorts to the process of neutron capture. Neutrons (due to their lack of charge) are readily absorbed by a nucleus. The heavy elements are created by either a *slow* neutron capture process (the so-called *s* process) or by the *rapid*, or *r* process. The *s* process occurs in thermally pulsing stars (called AGB, or asymptotic giant branch stars) and takes hundreds to thousands of years to reach the heaviest elements of lead and bismuth. The *r* process is thought to occur in supernova explosions because the conditions of high temperature, high neutron flux and ejected matter are present. These stellar conditions make the successive neutron captures very fast, involving very neutron-rich species which then beta-decay to heavier elements, especially at the so-called waiting points that correspond to more stable nuclides with closed neutron shells (magic numbers).

Condensed Matter Physics

Condensed matter physics deals with the physical properties of condensed phases of matter. These properties appear when a number of atoms at the supramolecular and macromolecular scale interact strongly and adhere to each other or are otherwise highly concentrated in a system. The most familiar examples of condensed phases are solids and liquids. Such every-day condensed phases arise from the electromagnetic forces between atoms. More exotic condensed phases include the mesophases of liquid crystal devices, the superconducting phase exhibited by certain materials at low temperature, the ferromagnetic and anti-ferromagnetic phases of spins on atomic lattices, and the Bose-Einstein condensate found in certain ultracold atomic systems.

Condensed matter physics seeks to understand the behaviour of these phases by using well-established physical laws. In particular, these include the laws of quantum mechanics, electromagnetism and statistical mechanics. The diversity of systems and phenomena available for study makes condensed matter physics by far the largest field of contemporary physics. By one estimate, one third of all United States physicists identify themselves as condensed matter physicists. The field has a large overlap with chemistry, materials science, and nanotechnology, and there are close connections with the related fields of atomic physics and biophysics. Theoretical condensed matter physics also shares many important concepts and techniques with theoretical particle and nuclear physics.

Historically, condensed matter physics grew out of solid-state physics, now considered one of its main subfields. The name of the field was apparently coined in 1967 by Philip Anderson and Volker Heine when they renamed their research group in the Cavendish

Laboratory of the University of Cambridge from "Solid-State Theory" to "Theory of Condensed Matter". In 1978, the Division of Solid State Physics at the American Physical Society was renamed as the Division of Condensed Matter Physics. One of the reasons for this change is that many of the concepts and techniques developed for studying solids can also be applied to fluid systems. For instance, the conduction electrons in an electrical conductor form a Fermi liquid, with similar properties to conventional liquids made up of atoms or molecules. Even the phenomenon of superconductivity, in which the quantum-mechanical properties of the electrons lead to collective behaviour fundamentally different from that of a classical fluid, is closely related to the superfluid phase of liquid helium.

Topics in Condensed Matter Physics

- Phases
  - *Generic phases* - gas (* uncondensed); liquid; solid
  - *Low temperature phases* - Fermi gas; Fermi liquid; Fermionic condensate; Luttinger liquid; superfluid; composite fermions; supersolid
  - *Phase phenomena* - order parameter; phase transition; cooling curve.
- Interfaces
  - *Surface tension*
  - *Domain growth* - nucleation; spinodal decomposition
  - *Interfacial growth* - dendritic growth; solidification fronts; viscous fingering
  - *Kapitza resistance* - liquid helium solid heat conductivity.
- crystalline solids
  - *Types* - insulator; metal; semiconductor; semimetal
  - *Electronic properties* - band gap; Bloch wave; conduction band; effective mass (solid-state physics); electrical conduction; electron hole; valence band
  - *Electronic phenomena* - Kondo effect; plasmon; quantum Hall effect; superconductivity; Wigner crystal; thermoelectricity
  - *Lattice phenomena* - anti-ferromagnet; ferroelectric effect; ferromagnet; magnon; phonon; spin glass; topological defect; multiferroics.

- Non-crystalline solids
  - *Types* - amorphous solid; granular matter; quasicrystals.
- Soft condensed matter
  - *Types* - liquid crystals; polymers; complex fluids; gels; foams; emulsions; colloids.
- Nanotechnology
  - *Nanoelectromechanical systems* (NEMS)
  - *Magnetic resonance force microscopy*
  - *Heat transport in nanoscale systems*
  - *Spin transport.*

5

Classical Mechanics

In physics, classical mechanics is one of the two major sub-fields of mechanics, which is concerned with the set of physical laws describing the motion of bodies under the action of a system of forces. The study of the motion of bodies is an ancient one, making classical mechanics one of the oldest and largest subjects in science, engineering and technology. Classical mechanics describes the motion of macroscopic objects, from projectiles to parts of machinery, as well as astronomical objects, such as spacecraft, planets, stars, and galaxies. Besides this, many specialisations within the subject deal with gases, liquids, solids, and other specific sub-topics. Classical mechanics provides extremely accurate results as long as the domain of study is restricted to large objects and the speeds involved do not approach the speed of light.

When the objects being dealt with become sufficiently small, it becomes necessary to introduce the other major sub-field of mechanics, quantum mechanics, which reconciles the macroscopic laws of physics with the atomic nature of matter and handles the wave-particle duality of atoms and molecules. In the case of high velocity objects approaching the speed of light, classical mechanics is enhanced by special relativity. General relativity unifies special relativity with Newton's law of universal gravitation, allowing physicists to handle gravitation at a deeper level.

The term *classical mechanics* was coined in the early 20th century to describe the system of physics begun by Isaac Newton and many contemporary 17th century natural philosophers, building upon the earlier astronomical theories of Johannes Kepler, which in turn were based on the precise observations of Tycho Brahe and the studies of terrestrial projectile motion of Galileo. Because these aspects of physics were developed long before the emergence of quantum physics and relativity, some sources exclude Einstein's theory of relativity from this category. However, a number of modern sources *do* include relativistic mechanics, which in their view represents *classical mechanics* in its most developed and most accurate form.

The initial stage in the development of classical mechanics is often referred to as Newtonian mechanics, and is associated with the physical concepts employed by and the mathematical methods invented by Newton himself, in parallel with Leibniz, and others. This is further described in the following sections.

Later, more abstract and general methods were developed, leading to reformulations of classical mechanics known as Lagrangian mechanics and Hamiltonian mechanics. These advances were largely made in the 18th and 19th centuries, and they extend substantially beyond Newton's work, particularly through their use of analytical mechanics. Ultimately, the mathematics developed for these were central to the creation of quantum mechanics.

Description of the Theory

The following introduces the basic concepts of classical mechanics. For simplicity, it often models real-world objects as point particles, objects with negligible size. The motion of a point particle is characterised by a small number of parameters: its position, mass, and the forces applied to it. Each of these parameters is discussed in turn. In reality, the kind of objects that classical mechanics can describe always have a non-zero size. (The physics of *very* small particles, such as the electron, is more accurately described by quantum mechanics). Objects with non-zero size have more complicated behaviour than hypothetical point particles, because of the additional degrees of freedom—for example, a baseball can spin while it is moving.

However, the results for point particles can be used to study such objects by treating them as composite objects, made up of a large number of interacting point particles. The center of mass of a composite object behaves like a point particle.

Position and its Derivatives

The *position* of a point particle is defined with respect to an arbitrary fixed reference point, O, in space, usually accompanied by a coordinate system, with the reference point located at the *origin* of the coordinate system. It is defined as the vector r from O to the particle. In general, the point particle need not be stationary relative to O, so r is a function of *t*, the time elapsed since an arbitrary initial time. In pre-Einstein relativity (known as Galilean relativity), time is considered an absolute, i.e., the time interval between any given pair of events is the same for all observers. In addition to relying on absolute time, classical mechanics assumes Euclidean geometry for the structure of space.

Table: *The SI derived "mechanical"(that is, not electromagnetic or thermal) units with kg, m and s*

Position	***m***
Angular position/Angle	unitless (radian)
velocity	$m\ s^{-1}$
Angular velocity	s^{-1}
acceleration	$m\ s^{-2}$
Angular acceleration	s^{-2}
jerk	$m\ s^{-3}$
"Angular jerk"	s^{-3}
specific energy	$m^2\ s^{-2}$
absorbed dose rate	$m^2\ s^{-3}$
moment of inertia	$kg\ m^2$
momentum	$kg\ m\ s^{-1}$
angular momentum	$kg\ m^2\ s^{-1}$
force	$kg\ m\ s^{-2}$
torque	$kg\ m^2\ s^{-2}$
energy	$kg\ m^2\ s^{-2}$
power	$kg\ m^2\ s^{-3}$
pressure and energy density	$kg\ m^{-1}\ s^{-2}$
surface tension	$kg\ s^{-2}$
Spring constant	$kg\ s^{-2}$
irradiance and energy flux	$kg\ s^{-3}$
kinematic viscosity	$m^2\ s^{-1}$
dynamic viscosity	$kg\ m^{-1}\ s^{-1}$
Density (mass density)	$kg\ m^{-3}$
Density (weight density)	$kg\ m^{-2}\ s^{-2}$
Number density	m^{-3}
Action	$kg\ m^2\ s^{-1}$

Velocity and Speed

The *velocity*, or the rate of change of position with time, is defined as the derivative of the position with respect to time or

$$v = \frac{dr}{dt},$$

In classical mechanics, velocities are directly additive and subtractive. For example, if one car travelling East at 60 km/h passes another car travelling East at 50 km/h, then from the perspective of the slower car, the faster car is travelling east at 60 H 50 = 10 km/h. Whereas, from the perspective of the faster car, the slower car is moving 10 km/h to the West. Velocities are directly additive as vector quantities; they must be dealt with using vector analysis.

Mathematically, if the velocity of the first object in the previous discussion is denoted by the vector u = *u*d and the velocity of the second object by the vector v = *v*e, where *u* is the speed of the first object, *v* is the speed of the second object, and d and e are unit vectors in the directions of motion of each particle respectively, then the velocity of the first object as seen by the second object is

$$u' = u - v.$$

Similarly,

$$v' = v - u.$$

When both objects are moving in the same direction, this equation can be simplified to

$$u' = (u - v)d.$$

Or, by ignoring direction, the difference can be given in terms of speed only:

$$u' = u - v.$$

Acceleration

The *acceleration*, or rate of change of velocity, is the derivative of the velocity with respect to time (the second derivative of the position with respect to time) or

$$a = \frac{dv}{dt}.$$

Acceleration can arise from a change with time of the magnitude of the velocity or of the direction of the velocity or both. If only the magnitude *v* of the velocity decreases, this is sometimes referred to as *deceleration*, but generally any change in the velocity with time, including deceleration, is simply referred to as acceleration.

Frames of Reference

While the position and velocity and acceleration of a particle can be referred to any observer in any state of motion, classical mechanics assumes the existence of a special family of reference frames in terms of which the mechanical laws of nature take a comparatively simple form. These special reference frames are called inertial frames. An inertial frame is such that when an object without any force interactions (an idealised situation) is viewed from it, it will appear either to be at rest or in a state of uniform motion in a straight line. This is the fundamental definition of an inertial frame. They are characterised by the requirement that all forces entering the observer's physical laws originate in identifiable sources (charges, gravitational bodies, and so forth). A non-inertial reference frame is one accelerating with respect to an inertial one, and in such a non-inertial frame a particle is subject to acceleration by fictitious forces that enter the equations of motion solely as a result of its accelerated motion, and do not originate in identifiable sources. These fictitious forces are in addition to the real forces recognised in an inertial frame. A key concept of inertial frames is the method for identifying them. For practical purposes, reference frames that are unaccelerated with respect to the distant stars are regarded as good approximations to inertial frames.

Consider two reference frames S and S'. For observers in each of the reference frames an event has space-time coordinates of (x,y,z,t) in frame S and (x', y', z', t') in frame S'. Assuming time is measured the same in all reference frames, and if we require $x = x'$ when $t = 0$, then the relation between the space-time coordinates of the same event observed from the reference frames $S2$ and S, which are moving at a relative velocity of u in the x direction is:

$$x' = x - ut$$

$$y' = y$$

$$z' = z$$

$$t' = t$$

This set of formulas defines a group transformation known as the Galilean transformation (informally, the *Galilean transform*). This group is a limiting case of the Poincaré group used in special relativity. The limiting case applies when the velocity u is very small compared to c, the speed of light.

The transformations have the following consequences:

- $v' = v - u$ (the velocity v' of a particle from the perspective of *S*2 is slower by u than its velocity v from the perspective of *S*)
- $a' = a$ (the acceleration of a particle is the same in any inertial reference frame)
- $F' = F$ (the force on a particle is the same in any inertial reference frame)
- the speed of light is not a constant in classical mechanics, nor does the special position given to the speed of light in relativistic mechanics have a counterpart in classical mechanics.

For some problems, it is convenient to use rotating coordinates (reference frames). Thereby one can either keep a mapping to a convenient inertial frame, or introduce additionally a fictitious centrifugal force and Coriolis force.

Forces; Newton's Second Law

Newton was the first to mathematically express the relationship between force and momentum. Some physicists interpret Newton's second law of motion as a definition of force and mass, while others consider it to be a fundamental postulate, a law of nature. Either interpretation has the same mathematical consequences, historically known as "Newton's Second Law":

$$F = \frac{dp}{dt} = \frac{d(mv)}{dt}.$$

The quantity mv is called the (canonical) momentum. The net force on a particle is thus equal to rate change of momentum of the particle with time. Since the definition of acceleration is $a = dv/dt$, the second law can be written in the simplified and more familiar form:

$$F = ma.$$

So long as the force acting on a particle is known, Newton's second law is sufficient to describe the motion of a particle. Once independent relations for each force acting on a particle are available, they can be substituted into Newton's second law to obtain an ordinary differential equation, which is called the *equation of motion*.

As an example, assume that friction is the only force acting on the particle, and that it may be modelled as a function of the velocity of the particle, for example:

$$F_R = -\lambda v,$$

where λ is a positive constant. Then the equation of motion is

$$-\lambda v = ma = m\frac{dv}{dt}.$$

This can be integrated to obtain

$$v = v_0 e^{-\lambda t/m}$$

where v_0 is the initial velocity. This means that the velocity of this particle decays exponentially to zero as time progresses. In this case, an equivalent viewpoint is that the kinetic energy of the particle is absorbed by friction (which converts it to heat energy in accordance with the conservation of energy), slowing it down. This expression can be further integrated to obtain the position r of the particle as a function of time.

Important forces include the gravitational force and the Lorentz force for electromagnetism. In addition, Newton's third law can sometimes be used to deduce the forces acting on a particle: if it is known that particle A exerts a force F on another particle B, it follows that B must exert an equal and opposite *reaction force*, "F, on A. The strong form of Newton's third law requires that F and "F act along the line connecting A and B, while the weak form does not. Illustrations of the weak form of Newton's third law are often found for magnetic forces.

Work and Energy

If a constant force F is applied to a particle that achieves a displacement Δr, the *work done* by the force is defined as the scalar product of the force and displacement vectors:

$$W = F.\Delta r.$$

More generally, if the force varies as a function of position as the particle moves from r_1 to r_2 along a path C, the work done on the particle is given by the line integral

$$W = \int_C F(r).dr.$$

If the work done in moving the particle from r_1 to r_2 is the same no matter what path is taken, the force is said to be conservative. Gravity is a conservative force, as is the force due to an idealised spring, as given by Hooke's law. The force due to friction is non-conservative.

The kinetic energy E_k of a particle of mass m travelling at speed v is given by

$$E_k = \frac{1}{2}mv^2.$$

For extended objects composed of many particles, the kinetic energy of the composite body is the sum of the kinetic energies of the particles.

The work-energy theorem states that for a particle of constant mass m the total work W done on the particle from position r_1 to r_2 is equal to the change in kinetic energy E_k of the particle:

$$W = \Delta E_k = E_{k,2} - E_{k,1} = \frac{1}{2}m\left(v_2^2 - v_1^2\right).$$

Conservative forces can be expressed as the gradient of a scalar function, known as the potential energy and denoted E_p:

$$F = -\nabla E_p.$$

If all the forces acting on a particle are conservative, and E_p is the total potential energy (which is defined as a work of involved forces to rearrange mutual positions of bodies), obtained by summing the potential energies corresponding to each force

$$F.\Delta r = -\Delta E_p.\Delta r = -\Delta E_p \Rightarrow -\Delta E_p = \Delta E_k \Rightarrow \Delta\left(E_k + E_p\right) = 0.$$

This result is known as *conservation of energy* and states that the total energy,

$$\sum E = E_k + E_p.$$

is constant in time. It is often useful, because many commonly encountered forces are conservative.

Beyond Newton's Laws

Classical mechanics also includes descriptions of the complex motions of extended non-pointlike objects. Euler's laws provide extensions to Newton's laws in this area. The concepts of angular momentum rely on the same calculus used to describe one-dimensional motion. The rocket equation extends the notion of rate of change of an object's momentum to include the effects of an object "losing mass".

There are two important alternative formulations of classical mechanics: Lagrangian mechanics and Hamiltonian mechanics. These, and other modern formulations, usually bypass the concept of "force", instead referring to other physical quantities, such as energy, for describing mechanical systems.

The expressions given above for momentum and kinetic energy are only valid when there is no significant electromagnetic contribution. In electromagnetism, Newton's second law for current-carrying wires breaks down unless one includes the electromagnetic field contribution to the momentum of the system as expressed by the Poynting vector divided by c^2, where c is the speed of light in free space.

History

Some Greek philosophers of antiquity, among them Aristotle, founder of Aristotelian physics, may have been the first to maintain the idea that "everything happens for a reason" and that theoretical principles can assist in the understanding of nature. While to a modern reader, many of these preserved ideas come forth as eminently reasonable, there is a conspicuous lack of both mathematical theory and controlled experiment, as we know it. These both turned out to be decisive factors in forming modern science, and they started out with classical mechanics.

The medieval "science of weights" (i.e., mechanics) owes much of its importance to the work of Jordanus de Nemore. In the *Elementa super demonstrationem ponderum*, he introduces the concept of "positional gravity" and the use of component forces.

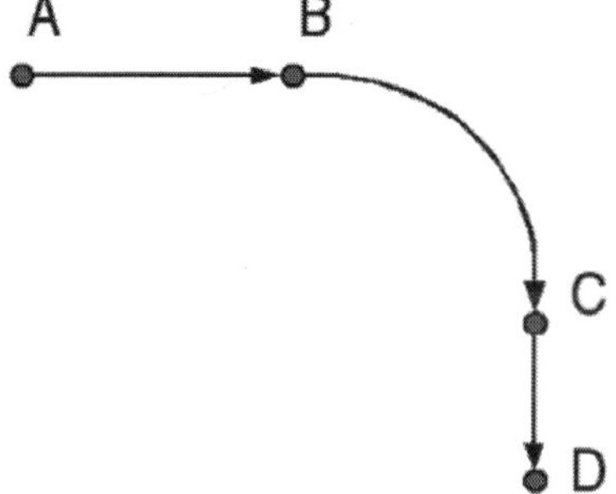

Figure: *Three stage Theory of impetus according to Albert of Saxony.*

The first published causal explanation of the motions of planets was Johannes Kepler's Astronomia nova published in 1609. He concluded, based on Tycho Brahe's observations of the orbit of Mars, that the orbits were ellipses. This break with ancient thought was happening around the same time that Galilei was proposing abstract mathematical laws for the motion of objects. He may (or may not) have performed the famous experiment of dropping two cannon balls of different weights from the tower of Pisa, showing that they both hit the ground at the same time. The reality of this experiment is disputed, but, more importantly, he did carry out quantitative experiments by rolling balls on an inclined plane. His theory of accelerated motion derived from the results of such experiments, and forms a cornerstone of classical mechanics.

As foundation for his principles of natural philosophy, Newton proposed three laws of motion: the law of inertia, his second law of acceleration, and the law of action and reaction; and hence laid the foundations for classical mechanics. Both Newton's second and third laws were given proper scientific and mathematical treatment in Newton's Philosophiæ Naturalis Principia Mathematica, which distinguishes them from earlier attempts at explaining similar phenomena, which were either incomplete, incorrect, or given little accurate mathematical expression. Newton also enunciated the principles of conservation of momentum and angular momentum. In Mechanics, Newton was also the first to provide the first correct scientific and mathematical formulation of gravity in Newton's law of universal gravitation. The combination of Newton's laws of motion and gravitation provide the fullest and most accurate description of classical mechanics. He demonstrated that these laws apply to everyday objects as well as to celestial objects. In particular, he obtained a theoretical explanation of Kepler's laws of motion of the planets.

Newton previously invented the calculus, of mathematics, and used it to perform the mathematical calculations. For acceptability, his book, the Principia, was formulated entirely in terms of the long established geometric methods, which were soon to be eclipsed by his calculus. However it was Leibniz who developed the notation of the derivative and integral preferred today.

Newton, and most of his contemporaries, with the notable exception of Huygens, worked on the assumption that classical mechanics would be able to explain all phenomena, including light, in the form of geometric optics. Even when discovering the so-called Newton's rings (a wave interference phenomenon) his explanation remained with his own corpuscular theory of light.

After Newton, classical mechanics became a principal field of study in mathematics as well as physics. After Newton there were several re-formulations which progressively allowed a solution to be found to a far greater number of problems. The first notable re-formulation was in 1788 by Joseph Louis Lagrange. Lagrangian mechanics was in turn re-formulated in 1833 by William Rowan Hamilton.

Some difficulties were discovered in the late 19th century that could only be resolved by more modern physics. Some of these difficulties related to compatibility with electromagnetic theory, and the famous Michelson-Morley experiment. The resolution of these problems led to the special theory of relativity, often included in the term classical mechanics.

A second set of difficulties were related to thermodynamics. When combined with thermodynamics, classical mechanics leads to the Gibbs paradox of classical statistical mechanics, in which entropy is not a well-defined quantity. Black-body radiation was not explained without the introduction of quanta. As experiments reached the atomic level, classical mechanics failed to explain, even approximately, such basic things as the energy levels and sizes of atoms and the photo-electric effect. The effort at resolving these problems led to the development of quantum mechanics.

Since the end of the 20th century, the place of classical mechanics in physics has been no longer that of an independent theory. Emphasis has shifted to understanding the fundamental forces of nature as in the Standard model and its more modern extensions into a unified theory of everything. Classical mechanics is a theory for the study of the motion of non-quantum mechanical, low-energy particles in weak gravitational fields. In the 21st century classical mechanics has been extended into the complex domain and complex classical mechanics exhibits behaviours very similar to quantum mechanics.

Limits of Validity

Many branches of classical mechanics are simplifications or approximations of more accurate forms; two of the most accurate being general relativity and relativistic statistical mechanics. Geometric optics is an approximation to the quantum theory of light, and does not have a superior "classical" form.

The Newtonian Approximation to Special Relativity

In special relativity, the momentum of a particle is given by

$$p = \frac{mv}{\sqrt{1 - \left(v^2 / c^2\right)}},$$

where m is the particle's mass, v its velocity, and c is the speed of light.

If v is very small compared to c, v^2/c^2 is approximately zero, and so

$$p \approx mv.$$

Thus the Newtonian equation p = mv is an approximation of the relativistic equation for bodies moving with low speeds compared to the speed of light.

For example, the relativistic cyclotron frequency of a cyclotron, gyrotron, or high voltage magnetron is given by

$$f = f_c \frac{m_0}{m_0 + T/c^2},$$

where f_c is the classical frequency of an electron (or other charged particle) with kinetic energy T and (rest) mass m_0 circling in a magnetic field. The (rest) mass of an electron is 511 keV. So the frequency correction is 1% for a magnetic vacuum tube with a 5.11 kV direct current accelerating voltage.

The Classical Approximation to Quantum Mechanics

The ray approximation of classical mechanics breaks down when the de Broglie wavelength is not much smaller than other dimensions of the system. For non-relativistic particles, this wavelength is.

$$\lambda = \frac{h}{p}$$

where h is Planck's constant and p is the momentum.

Again, this happens with electrons before it happens with heavier particles. For example, the electrons used by Clinton Davisson and Lester Germer in 1927, accelerated by 54 volts, had a wave length of 0.167 nm, which was long enough to exhibit a single diffraction side lobe when reflecting from the face of a nickel crystal with atomic spacing of 0.215 nm. With a larger vacuum chamber, it would seem relatively easy to increase the angular resolution from around a radian to a milliradian and see quantum diffraction from the periodic patterns of integrated circuit computer memory.

More practical examples of the failure of classical mechanics on an engineering scale are conduction by quantum tunneling in tunnel diodes and very narrow transistor gates in integrated circuits.

Classical mechanics is the same extreme high frequency approximation as geometric optics. It is more often accurate because it describes particles and bodies with rest mass. These have more momentum and therefore shorter De Broglie wavelengths than massless particles, such as light, with the same kinetic energies.

Branches

Classical mechanics was traditionally divided into three main branches:

- Statics, the study of equilibrium and its relation to forces
- Dynamics, the study of motion and its relation to forces
- Kinematics, dealing with the implications of observed motions without regard for circumstances causing them.

Another division is based on the choice of mathematical formalism:

- Newtonian mechanics
- Lagrangian mechanics
- Hamiltonian mechanics.

Alternatively, a division can be made by region of application:

- Celestial mechanics, relating to stars, planets and other celestial bodies
- Continuum mechanics, for materials which are modelled as a continuum, e.g., solids and fluids (i.e., liquids and gases).
- Relativistic mechanics (i.e. including the special and general theories of relativity), for bodies whose speed is close to the speed of light.
- Statistical mechanics, which provides a framework for relating the microscopic properties of individual atoms and molecules to the macroscopic or bulk thermodynamic properties of materials.

Hamiltonian Mechanics

Hamiltonian mechanics is a reformulation of classical mechanics that was introduced in 1833 by Irish mathematician William Rowan Hamilton.

It arose from Lagrangian mechanics, a previous reformulation of classical mechanics introduced by Joseph Louis Lagrange in 1788, but can be formulated *without* recourse to Lagrangian mechanics using symplectic spaces. The Hamiltonian method differs from the Lagrangian method in that instead of expressing second-order differential constraints on an n-dimensional coordinate space (where n is the number of degrees of freedom of the system), it expresses first-order constraints on a $2n$-dimensional phase space.

As with Lagrangian mechanics, Hamilton's equations provide a new and equivalent way of looking at classical mechanics. Generally, these equations do not provide a more convenient way of solving a particular problem. Rather, they provide deeper insights into both the general structure of classical mechanics and its connection to quantum mechanics as understood through Hamiltonian mechanics, as well as its connection to other areas of science.

Simplified Overview of Uses

The value of the Hamiltonian is the total energy of the system being described. For a closed system, it is the sum of the kinetic and potential energy in the system. There is a set of differential equations known as the *Hamilton equations* which give the time evolution of the system. Hamiltonians can be used to describe such simple systems as a bouncing ball, a pendulum or an oscillating spring in which energy changes from kinetic to potential and back again over time. Hamiltonians can also be employed to model the energy of other more complex dynamic systems such as planetary orbits in celestial mechanics and also in quantum mechanics.

The Hamilton equations are generally written as follows:

$$\dot{p} = -\frac{\partial H}{\partial q}$$

$$\dot{q} = -\frac{\partial H}{\partial p}.$$

In the above equations, the dot denotes the ordinary derivative with respect to time of the functions $p = p(t)$ (called generalised momenta) and $q = q(t)$ (called generalised coordinates), taking values in some vector space, and H=H(p, q, t) is the so-called Hamiltonian, or (scalar valued) Hamiltonian function. Thus, more explicitly, one can equivalently write

$$\frac{d}{dt}p(t) = -\frac{\partial}{\partial q}H(p(t), q(t), t)$$

$$\frac{d}{dt}q(t) = -\frac{\partial}{\partial p}H(p(t), q(t), t)$$

and specify the domain of values in which the parameter t (time) varies.

For a detailed derivation of these equations from Lagrangian mechanics.

Basic Physical Interpretation

The simplest interpretation of the Hamilton equations is as follows, applying them to a one-dimensional system consisting of one particle of mass m under time-independent boundary conditions: The Hamiltonian H represents the energy of the system (provided that there are NO external forces, or additional energy added to the system), which is the sum of kinetic and potential energy, traditionally denoted T and V, respectively. Here q is the x coordinate and p is the momentum, mv. Then

$$H = T + V,\ T = \frac{p^2}{2m},\ V = V(q) = V(x).$$

Note that T is a function of p alone, while V is a function of x (or q) alone.

Now the time-derivative of the momentum p equals the *Newtonian force*, and so here the first Hamilton equation means that the force on the particle equals the rate at which it loses potential energy with respect to changes in x, its location. (Force equals the negative gradient of potential energy.) The time-derivative of q here means the velocity: the second Hamilton equation here means that the particle's velocity equals the derivative of its kinetic energy with respect to its momentum. (Because the derivative with respect to p of $p^2/2m$ equals $p/m = mv/m = v$.)

Geometry of Hamiltonian Systems

A Hamiltonian system may be understood as a fibre bundle E over time R, with the fibres E_t, $t - R$ being the position space. The Lagrangian is thus a function on the jet bundle J over E; taking the fibrewise Legendre transform of the Lagrangian produces a function on the dual bundle over time whose fibre at t is the cotangent space T^*E_t, which comes equipped with a natural symplectic form, and this latter function is the Hamiltonian.

Generalisation to Quantum Mechanics Through Poisson Bracket

Hamilton's equations above work well for classical mechanics, but not for quantum mechanics, since the differential equations discussed assume that one can specify the exact position and momentum of the particle simultaneously at any point in time. However, the equations can be further generalised to then be extended to apply to quantum mechanics as well as to classical mechanics, through the deformation of the Poisson algebra over p and q to the algebra of Moyal brackets.

Specifically, the more general form of the Hamilton's equation reads

$$\frac{df}{dt} = \{f, H\} + \frac{\partial f}{\partial t}$$

where f is some function of p and q, and H is the Hamiltonian. To find out the rules for evaluating a Poisson bracket without resorting to differential equations, a Poisson bracket is the name for the Lie bracket in a Poisson algebra. These Poisson brackets can then be extended to Moyal brackets comporting to an inequivalent Lie algebra, as proven by H Groenewold, and thereby describe quantum mechanical diffusion in phase space. This more algebraic approach not only permits ultimately extending probability distributions in phase space to Wigner quasi-probability distributions, but, at the mere Poisson bracket classical setting, also provides more power in helping analyse the relevant conserved quantities in a system.

Mathematical Formalism

Any smooth real-valued function H on a symplectic manifold can be used to define a Hamiltonian system. The function H is known as the Hamiltonian or the energy function. The symplectic manifold is then called the phase space. The Hamiltonian induces a special vector field on the symplectic manifold, known as the symplectic vector field.

The symplectic vector field, also called the Hamiltonian vector field, induces a Hamiltonian flow on the manifold. The integral curves of the vector field are a one-parameter family of transformations of the manifold; the parameter of the curves is commonly called the time. The time evolution is given by symplectomorphisms. By Liouville's theorem, each symplectomorphism preserves the volume form on the phase space. The collection of symplectomorphisms induced by the Hamiltonian flow is commonly called the Hamiltonian mechanics of the Hamiltonian system.

The symplectic structure induces a Poisson bracket. The Poisson bracket gives the space of functions on the manifold the structure of a Lie algebra.

Given a function f

$$\frac{df}{dt}f = \frac{\partial}{\partial t}f + \{f, H\}.$$

If we have a probability distribution, ρ, then (since the phase space velocity $(\dot{p}_i, \dot{q}_i)$ has zero divergence, and probability is conserved) its convective derivative can be shown to be zero and so

$$\frac{\partial}{\partial t}\rho = -\{p, H\}.$$

This is called Liouville's theorem. Every smooth function G over the symplectic manifold generates a one-parameter family of symplectomorphisms and if $\{G, H\} = 0$, then G is conserved and the symplectomorphisms are symmetry transformations.

A Hamiltonian may have multiple conserved quantities G_i. If the symplectic manifold has dimension $2n$ and there are n functionally independent conserved quantities G_i which are in involution (i.e., $\{G_i, G_j\} = 0$), then the Hamiltonian is Liouville integrable. The Liouville–Arnol'd theorem says that locally, any Liouville integrable Hamiltonian can be transformed via a symplectomorphism in a new Hamiltonian with the conserved quantities G_i as coordinates; the new coordinates are called *action-angle coordinates*. The transformed Hamiltonian depends only on the G_i, and hence the equations of motion have the simple form

$$\dot{G}_i = 0, \ \dot{\varphi}_i = G(G),$$

for some function F (Arnol'd et al., 1988). There is an entire field focusing on small deviations from integrable systems governed by the KAM theorem.

The integrability of Hamiltonian vector fields is an open question. In general, Hamiltonian systems are chaotic; concepts of measure, completeness, integrability and stability are poorly defined. At this time, the study of dynamical systems is primarily qualitative, and not a quantitative science.

Riemannian Manifolds

An important special case consists of those Hamiltonians that are quadratic forms, that is, Hamiltonians that can be written as

$$H(q,p) = \frac{1}{2}\langle p,p\rangle_q$$

where $\langle \cdot,\cdot \rangle_q$ is a smoothly varying inner product on the fibres T_q^*Q, the cotangent space to the point q in the configuration space, sometimes called a cometric. This Hamiltonian consists entirely of the kinetic term.

If one considers a Riemannian manifold or a pseudo-Riemannian manifold, the Riemannian metric induces a linear isomorphism between the tangent and cotangent bundles. Using this isomorphism, one can define a cometric. (In coordinates, the matrix defining the cometric is the inverse of the matrix defining the metric.) The solutions to the Hamilton–Jacobi equations for this Hamiltonian are then the same as the geodesics on the manifold. In particular, the Hamiltonian flow in this case is the same thing as the geodesic flow. The existence of such solutions, and the completeness of the set of solutions.

Sub-Riemannian Manifolds

When the cometric is degenerate, then it is not invertible. In this case, one does not have a Riemannian manifold, as one does not have a metric. However, the Hamiltonian still exists. In the case where the cometric is degenerate at every point q of the configuration space manifold Q, so that the rank of the cometric is less than the dimension of the manifold Q, one has a sub-Riemannian manifold.

The Hamiltonian in this case is known as a sub-Riemannian Hamiltonian. Every such Hamiltonian uniquely determines the cometric, and vice-versa. This implies that every sub-Riemannian manifold is uniquely determined by its sub-Riemannian Hamiltonian, and that the converse is true: every sub-Riemannian manifold has a unique sub-Riemannian Hamiltonian. The existence of sub-Riemannian geodesics is given by the Chow-Rashevskii theorem.

The continuous, real-valued Heisenberg group provides a simple example of a sub-Riemannian manifold. For the Heisenberg group, the Hamiltonian is given by

$$H(x, y, z, p_x, p_y, p_z) = \frac{1}{2}\left(p_x^2 + p_y^2\right).$$

p_z is not involved in the Hamiltonian.

Poisson Algebras

Hamiltonian systems can be generalised in various ways. Instead of simply looking at the algebra of smooth functions over a symplectic manifold, Hamiltonian mechanics can be formulated on general commutative unital real Poisson algebras. A state is a continuous linear functional on the Poisson algebra (equipped with some suitable topology) such that for any element A of the algebra, A^2 maps to a non-negative real number.

A further generalisation is given by Nambu dynamics.

Celestial Mechanics

Celestial mechanics is the branch of astronomy that deals with the motions of celestial objects. The field applies principles of physics, historically classical mechanics, to astronomical objects such as stars and planets to produce ephemeris data. Orbital mechanics (astrodynamics) is a subfield which focuses on the orbits of artificial satellites. Lunar theory is another subfield focusing on the orbit of the Moon.

History of Celestial Mechanics

Modern analytic celestial mechanics started over 300 years ago with Isaac Newton's Principia of 1687. The name "celestial mechanics" is more recent than that. Newton wrote that the field should be called "rational mechanics." The term "dynamics" came in a little later with Gottfried Leibniz, and over a century after Newton, Pierre-Simon Laplace introduced the term "celestial mechanics." Nevertheless, prior studies addressing the problem of planetary positions are known going back perhaps 3,000 or more years, as early as the Babylonian astronomers.

Classical Greek writers speculated widely regarding celestial motions, and presented many geometrical mechanisms to model the motions of the planets. Their models employed combinations of uniform circular motion and were centred on the earth. An independent philosophical tradition was concerned with the physical causes of such circular motions. An extraordinary figure among the ancient Greek astronomers is Aristarchus of Samos (310 BCE–c.230 BCE), who suggested a heliocentric model of the universe and attempted to measure Earth's distance from the Sun.

The only known supporter of Aristarchus was Seleucus of Seleucia, a Babylonian astronomer who is said to have proved heliocentrism through reasoning in the 2nd century BCE. This may have involved the phenomenon of tides, which he correctly theorised to be caused by attraction to the Moon and notes that the height of the tides depends on the Moon's position relative to the Sun. Alternatively, he may have determined the constants of a geometric model for the heliocentric theory and developed methods to compute planetary positions using this model, possibly using early trigonometric methods that were available in his time, much like Copernicus.

Claudius Ptolemy

Claudius Ptolemy (c.120 CE) was an ancient astronomer and astrologer in early Imperial Roman times who wrote several books on astronomy. The most significant of these was the *Almagest*, which remained the most important book on predictive geometrical astronomy for some 1400 years. Ptolemy selected the best of the astronomical principles of his Greek predecessors, especially Hipparchus, and appears to have combined them either directly or indirectly with data and parameters obtained from the Babylonians. Although Ptolemy relied mainly on the work of Hipparchus, he introduced at least one idea, the equant, which appears to be his own, and which greatly improved the accuracy of the predicted positions of the planets. Although his model was extremely accurate, it relied solely on geometrical constructions rather than on physical causes; Ptolemy did not use celestial mechanics.

Early Middle Ages

B. L. van der Waerden has interpreted the planetary models developed by Aryabhata (476–550 CE), an Indian astronomer, and Albumasar (787–886 CE), a Persian astronomer, to be heliocentric models but this view has been strongly disputed by others. In the 9th century CE, the Persian physicist and astronomer, Ja'far Muhammad ibn Musa ibn Shakir, hypothesised that the heavenly bodies and celestial spheres are subject to the same laws of physics as Earth, unlike the ancients who believed that the celestial spheres followed their own set of physical laws different from that of Earth.

Ibn al-Haytham

In the early 11th century CE, Ibn al-Haytham presented a development of Ptolemy's geocentric epicyclic models in terms of nested celestial spheres.

Late Middle Ages

There was much debate on the dynamics of the celestial spheres during the late Middle Ages. Averroes (Ibn Rushd), Ibn Bajjah (Avempace) and Thomas Aquinas developed the theory of inertia in the celestial spheres, while Avicenna (Ibn Sina) and Jean Buridan developed the theory of impetus in the celestial spheres.

In the 14th century, Ibn al-Shatir produced the first model of lunar motion which matched physical observations, and which was later used by Copernicus. In the 13th–15th centuries, Tusi and Ali Ku°çu provided the earliest empirical evidence for the Earth's rotation, using the phenomena of comets to refute Ptolemy's claim that a stationary Earth can be determined through observation. Ku°çu further rejected Aristotelian physics and natural philosophy, allowing astronomy and physics to become empirical and mathematical instead of philosophical. In the early 16th century, the debate on the Earth's motion was continued by Al-Birjandi (d. 1528), who in his analysis of what might occur if the Earth were rotating, develops a hypothesis similar to Galileo Galilei's notion of "circular inertia", which he described in the following observational test:

"The small or large rock will fall to the Earth along the path of a line that is perpendicular to the plane (*sath*) of the horizon; this is witnessed by experience (*tajriba*). And this perpendicular is away from the tangent point of the Earth's sphere and the plane of the perceived (*hissi*) horizon. This point moves with the motion of the Earth and thus there will be no difference in place of fall of the two rocks."

Johannes Kepler

Johannes Kepler (27 December 1571–15 November 1630) was the first to closely integrate the predictive geometrical astronomy, which had been dominant from Ptolemy to Copernicus, with physical concepts to produce a *New Astronomy, Based upon Causes, or Celestial Physics....* His work led to the modern laws of planetary orbits, which he developed using his physical principles and the planetary observations made by Tycho Brahe. Kepler's model greatly improved the accuracy of predictions of planetary motion, years before Isaac Newton developed his law of gravitation.

Isaac Newton

Isaac Newton (4 January 1643–31 March 1727) is credited with introducing the idea that the motion of objects in the heavens, such as planets, the Sun, and the Moon, and the motion of objects on the ground, like cannon balls and falling apples, could be described by the same set of physical laws. In this sense he unified *celestial* and *terrestrial* dynamics. Using Newton's law of universal gravitation, proving Kepler's Laws for the case of a circular orbit is simple. Elliptical orbits involve more complex calculations, which Newton included in his Principia.

Joseph-Louis Lagrange

After Newton, Lagrange (25 January 1736–10 April 1813) attempted to solve the three-body problem, analysed the stability of planetary orbits, and discovered the existence of the Lagrangian points. Lagrange also reformulated the principles of classical mechanics, emphasising energy more than force and developing a method to use a single polar coordinate equation to describe any orbit, even those that are parabolic and hyperbolic. This is useful for calculating the behaviour of planets and comets and such. More recently, it has also become useful to calculate spacecraft trajectories.

Simon Newcomb

Simon Newcomb (12 March 1835–11 July 1909) was a Canadian-American astronomer who revised Peter Andreas Hansen's table of lunar positions. In 1877, assisted by George William Hill, he recalculated all the major astronomical constants. After 1884, he conceived with A. M. W. Downing a plan to resolve much international confusion on the subject. By the time he attended a standardisation conference in Paris, France in May 1886, the international consensus was that all ephemerides should be based on Newcomb's calculations. A further conference as late as 1950 confirmed Newcomb's constants as the international standard.

Albert Einstein

Albert Einstein (14 March 1879–18 April 1955) explained the anomalous precession of Mercury's perihelion in his 1916 paper *The Foundation of the General Theory of Relativity*. This led astronomers to recognise that Newtonian mechanics did not provide the highest accuracy. Binary pulsars have been observed, the first in 1974, whose orbits not only require the use of General Relativity for their explanation, but whose evolution proves the existence of gravitational radiation, a discovery that led to the 1993 Nobel Physics Prize.

Examples of Problems

Celestial motion without additional forces such as thrust of a rocket, is governed by gravitational acceleration of masses due to other masses. A simplification is the n-body problem, where the problem assumes some number n of spherically symmetric masses. In that case, the integration of the accelerations can be well approximated by relatively simple summations.

Examples:

- 4-body problem: spaceflight to Mars (for parts of the flight the influence of one or two bodies is very small, so that there we have a 2- or 3-body problem.
- 3-body problem:
  - — Quasi-satellite
  - — Spaceflight to, and stay at a Lagrangian point

In the case that n=2 (two-body problem), the situation is much simpler than for larger n. Various explicit formulas apply, where in the more general case typically only numerical solutions are possible. It is a useful simplification that is often approximately valid.

Examples:

- A binary star, e.g., Alpha Centauri (approx. the same mass)
- A binary asteroid, e.g., 90 Antiope (approx. the same mass)

A further simplification is based on the "standard assumptions in astrodynamics", which include that one body, the orbiting body, is much smaller than the other, the central body. This is also often approximately valid.

Examples:

- Solar system orbiting the center of the Milky Way
- A planet orbiting the Sun
- A moon orbiting a planet
- A spacecraft orbiting Earth, a moon, or a planet (in the latter cases the approximation only applies after arrival at that orbit)

Either instead of, or on top of the previous simplification, we may assume circular orbits, making distance and orbital speeds, and potential and kinetic energies constant in time. This assumption sacrifices accuracy for simplicity, especially for high eccentricity orbits which are by definition non-circular.

Examples:

- The orbit of the dwarf planet Pluto, ecc. = 0.2488
- The orbit of Mercury, ecc. = 0.2056
- Hohmann transfer orbit
- Gemini 11 flight
- Suborbital flights.

Perturbation Theory

Perturbation theory comprises mathematical methods that are used to find an approximate solution to a problem which cannot be solved exactly. (It is closely related to methods used in numerical analysis, which are ancient.) The earliest use of perturbation theory was to deal with the otherwise unsolveable mathematical problems of celestial mechanics: Newton's solution for the orbit of the Moon, which moves noticeably differently from a simple Keplerian ellipse because of the competing gravitation of the Earth and the Sun.

Perturbation methods start with a simplified form of the original problem, which is carefully chosen to be exactly solvable. In celestial mechanics, this is usually a Keplerian ellipse, which is correct when there are only two gravitating bodies (say, the Earth and the Moon), or a circular orbit, which is only correct in special cases of two-body motion, but is often close enough for practical use.

The solved, but simplified problem is then *"perturbed"* to make its starting conditions closer to the real problem, such as including the gravitational attraction of a third body (the Sun). The slight changes that result, which themselves may have been simplified yet again, are used as corrections. Because of simplifications introduced along every step of the way, the corrections are never perfect, but even one cycle of corrections often provides a remarkably better approximate solution to the real problem.

There is no requirement to stop at only one cycle of corrections. A partially corrected solution can be re-used as the new starting point for yet another cycle of perturbations and corrections. The common difficulty with the method is that usually the corrections progressively make the new solutions very much more complicated, so each cycle is much more difficult to manage than the previous cycle of corrections. Newton is reported to have said, regarding the problem of the Moon's orbit *"It causeth my head to ache."*

This general procedure – starting with a simplified problem and gradually adding corrections that make the starting point of the corrected problem closer to the real situation – is a widely used mathematical tool in advanced sciences and engineering. It is the natural extension of the "guess, check, and fix" method used anciently with numbers.

Orbital Mechanics

Orbital mechanics or astrodynamics is the application of ballistics and celestial mechanics to the practical problems concerning the motion of rockets and other spacecraft. The motion of these objects is usually calculated from Newton's laws of motion and Newton's law of universal gravitation. It is a core discipline within space mission design and control. Celestial mechanics treats more broadly the orbital dynamics of systems under the influence of gravity, including both spacecraft and natural astronomical bodies such as star systems, planets, moons, and comets. Orbital mechanics focuses on spacecraft trajectories, including orbital maneuvers, orbit plane changes, and interplanetary transfers, and is used by mission planners to predict the results of propulsive maneuvers. General relativity is a more exact theory than Newton's laws for calculating orbits, and is sometimes necessary for greater accuracy or in high-gravity situations (such as orbits close to the Sun).

History

Until the rise of space travel in the twentieth century, there was little distinction between orbital and celestial mechanics. The fundamental techniques, such as those used to solve the Keplerian problem (determining position as a function of time), are therefore the same in both fields. Furthermore, the history of the fields is almost entirely shared.

Johannes Kepler was the first to successfully model planetary orbits to a high degree of accuracy, publishing his laws in 1605. Isaac Newton published more general laws of celestial motion in his 1687 book, *Philosophiæ Naturalis Principia Mathematica.*

Practical Techniques

Rules of Thumb

The following rules of thumb are useful for situations approximated by classical mechanics under the standard assumptions of astrodynamics. The specific example discussed is of a satellite orbiting a planet, but the rules of thumb could also apply to other situations, such as orbits of small bodies around a star such as the Sun.

- Kepler's laws of planetary motion, which can be mathematically derived from Newton's laws, hold strictly only in describing the motion of two gravitating bodies, in the absence of non-gravitational forces, or approximately when the gravity of a single massive body like the Sun dominates other effects:
  - Orbits are elliptical, with the planet at one focus of the ellipse. Special cases of this are circular orbits (a circle being simply an ellipse of zero eccentricity) with the planet at the center, and parabolic orbits (which are ellipses with eccentricity of exactly 1, which is simply an infinitely long ellipse) with the planet at the focus.
  - A line drawn from the planet to the satellite sweeps out *equal areas in equal times* no matter which portion of the orbit is measured.
  - The square of a satellite's orbital period is proportional to the cube of its average distance from the planet.
- Without applying thrust (such as firing a rocket engine), the height and shape of the satellite's orbit won't change, and it will maintain the same orientation with respect to the fixed stars.
- A satellite in a low orbit (or low part of an elliptical orbit) moves more quickly with respect to the surface of the planet than a satellite in a higher orbit (or a high part of an elliptical

orbit), duc to the stronger gravitational attraction closer to the planet.

- If thrust is applied at only one point in the satellite's orbit, it will return to that same point on each subsequent orbit, though the rest of its path will change. Thus to move from one circular orbit to another, at least two brief applications of thrust are needed.
- From a circular orbit, thrust in a direction which slows the satellite down will create an elliptical orbit with a lower periapse (lowest orbital point) at 180 degrees away from the firing point. If thrust is applied to speed the satellite, it will create an elliptical orbit with a higher apoapse 180 degrees away from the firing point.

The consequences of the rules of orbital mechanics are sometimes counter-intuitive. For example, if two spacecraft are in the same circular orbit and wish to dock, unless they are very close, the trailing craft cannot simply fire its engines to go faster. This will change the shape of its orbit, causing it to gain altitude and miss its target. One approach is to actually fire a reverse thrust to slow down, and then fire again to re-circularise the orbit at a lower altitude. Because lower orbits are faster than higher orbits, the trailing craft will begin to catch up. A third firing at the right time will put the trailing craft in an elliptical orbit which will intersect the path of the leading craft, approaching from below.

To the degree that the standard assumptions of astrodynamics do not hold, actual trajectories will vary from those calculated. For example, simple atmospheric drag is another complicating factor for objects in Earth orbit. These rules of thumb are decidedly inaccurate when describing two or more bodies of similar mass, such as a binary star system. (Celestial mechanics uses more general rules applicable to a wider variety of situations.) The differences between classical mechanics and general relativity can also become important for large objects like planets.

Orbital Transfer

Transfer orbits are usually elliptical orbits that allow spacecraft to move from one (usually substantially circular) orbit to another. Usually they require a burn at the start, a burn at the end, and sometimes one or more burns in the middle.

- Hohmann transfer orbit requires little delta-v.

- Bi-elliptic transfers sometimes require less.
- For faster transfers, any orbit that intersects both the original orbit and destination orbit may be used.

Gravity Assist and the Oberth Effect

In a gravity assist, a spacecraft swings by a planet and leaves in a different direction, at a different speed. This is useful to speed or slow a spacecraft instead of carrying more fuel.

This maneuver can be approximated by an elastic collision at large distances, though the flyby does not involve any physical contact. Due to Newton's Third Law (equal and opposite reaction), any momentum gained by a spacecraft must be lost by the planet, or vice versa. However, because the planet is much, much more massive than the spacecraft, the effect on the planet's orbit is negligible.

The Oberth effect can be employed, particularly during a gravity assist operation. This effect is that use of a propulsion system works better at high speeds, and hence course changes are best done when close to a gravitating body; this can multiply the effective delta-v.

Interplanetary Transport Network and fuzzy orbits

It is now possible to use computers to search for routes using the non-linearities in the gravity of the planets and moons of the solar system. For example, it is possible to plot an orbit from high earth orbit to Mars, passing close to one of the Earth's Trojan points. Collectively referred to as the Interplanetary Transport Network, these highly perturbative, even chaotic, orbital trajectories in principle need no fuel (in practice keeping to the trajectory requires some course corrections). The biggest problem with them is they can be exceedingly slow, taking many years to arrive. In addition launch windows can be very far apart.

They have, however, been employed on projects such as Genesis. This spacecraft visited Earth's Lagrange L_1 point and returned using very little propellant.

Laws of Astrodynamics

The fundamental laws of astrodynamics are Newton's law of universal gravitation and Newton's laws of motion, while the fundamental mathematical tool is his differential calculus.

Every orbit and trajectory outside atmospheres is in principle reversible, i.e., in the space-time function the time is reversed. The velocities are reversed and the accelerations are the same, including those due to rocket bursts. Thus if a rocket burst is in the direction of the velocity, in the reversed case it is opposite to the velocity. Of course in the case of rocket bursts there is no full reversal of events, both ways the same delta-v is used and the same mass ratio applies.

Standard assumptions in astrodynamics include non-interference from outside bodies, negligible mass for one of the bodies, and negligible other forces (such as from the solar wind, atmospheric drag, etc.). More accurate calculations can be made without these simplifying assumptions, but they are more complicated. The increased accuracy often does not make enough of a difference in the calculation to be worthwhile.

Kepler's laws of planetary motion may be derived from Newton's laws, when it is assumed that the orbiting body is subject only to the gravitational force of the central attractor. When an engine thrust or propulsive force is present, Newton's laws still apply, but Kepler's laws are invalidated. When the thrust stops, the resulting orbit will be different but will once again be described by Kepler's laws. The three laws are:

1. The orbit of every planet is an ellipse with the sun at one of the foci.
2. A line joining a planet and the sun sweeps out equal areas during equal intervals of time.
3. The squares of the orbital periods of planets are directly proportional to the cubes of the semi-major axis of the orbits.

Escape Velocity

The formula for escape velocity is easily derived as follows. The specific energy (energy per unit mass) of any space vehicle is composed of two components, the specific potential energy and the specific kinetic energy. The specific potential energy associated with a planet of mass *M* is given by

$$-GM/r$$

while the specific kinetic energy of an object is given by

$$v^2/2$$

Since energy is conserved, the total specific orbital energy

$$v^2/2 - GM/r$$

does not depend on the distance, *r*, from the center of the central body to the space vehicle in question. Therefore, the object can reach infinite *r* only if this quantity is non-negative, which implies

$$v \geq \sqrt{2GM/r}$$

The escape velocity from the Earth's surface is about 11 km/s, but that is insufficient to send the body an infinite distance because of the gravitational pull of the Sun. To escape the solar system from a location at a distance from the Sun equal to the distance Sun-Earth, but not close to the Earth, requires around 42 km/s velocity, but there will be "part credit" for the Earth's orbital velocity for spacecraft launched from Earth, if their further acceleration (due to the propulsion system) carries them in the same direction as Earth travels in its orbit.

Formulae For Free Orbits

Orbits are conic sections, so, naturally, the formulas for the distance of a body for a given angle corresponds to the formula for that curve in polar coordinates, which is:

$$r = \frac{p}{1 + e\cos\theta}$$

$$\mu = G(m_1 + m_2)$$

$$p = h^2/\mu$$

where μ is called the gravitational parameter, G is the gravitational constant, m_1 and m_2 are the masses of objects 1 and 2, and h is the angular momentum of object 2 with respect to object 1. The parameter θ is known as the true anomaly, p is the semi-latus rectum, while e is the orbital eccentricity, all obtainable from the various forms of the six independent orbital elements.

Circular Orbits

All bounded orbits where the gravity of a central body dominates are elliptical in nature. A special case of this is the circular orbit, which is an ellipse of zero eccentricity. The formula for the velocity of a body in a circular orbit at distance r from the center of gravity of mass M is

$$v = \sqrt{\frac{GM}{r}}$$

where G is the gravitational constant, equal to

$$6.672\ 598 \times 10^{-11}\ \text{m}^3/(\text{kg}\cdot\text{s}^2)$$

To properly use this formula, the units must be consistent; for example, M must be in kilograms, and r must be in meters. The answer will be in meters per second.

The quantity *GM* is often termed the standard gravitational parameter, which has a different value for every planet or moon in the solar system.

Once the circular orbital velocity is known, the escape velocity is easily found by multiplying by the square root of 2:

$$v = \sqrt{2}\sqrt{\frac{GM}{r}} = \sqrt{\frac{2GM}{r}}.$$

Elliptical Orbits

If 0<e<1, then the denominator of the equation of free orbits varies with the true anomaly θ, but remains positive, never becoming zero. Therefore, the relative position vector remains bounded, having its smallest magnitude at periapsis r_p which is given by:

$$r_p = \frac{p}{1+e}$$

The maximum value r is reached when θ = 180. This point is called the apoapsis, and its radial coordinate, denoted r_a, is

$$r_a = \frac{p}{1-e}$$

Let 2a be the distance measured along the apse line from periapsis P to apoapsis A, as illustrated in the equation below:

$$a = r_p + r_a$$

Substituting the equations above, we get:

$$a = \frac{p}{1-e^2}$$

a is the semimajor axis of the ellipse. Solving for r we get:

$$r = \frac{a\left(1-e^2\right)}{1+e\cos\theta}$$

Orbital Period

Under standard assumptions the orbital period (T) of a body travelling along an elliptic orbit can be computed as:

$$T = 2\pi\sqrt{\frac{a^2}{\mu}}$$

where:

- is standard gravitational parameter,
- is length of semi-major axis.

Conclusions:

- The orbital period is equal to that for a circular orbit with the orbit radius equal to the semi-major axis (a),
- For a given semi-major axis the orbital period does not depend on the eccentricity.

Velocity

Under standard assumptions the orbital speed (v) of a body travelling along elliptic orbit can be computed from the Vis-viva equation as:

$$v = \sqrt{\mu\left(\frac{2}{r} - \frac{1}{a}\right)}$$

where:

- μ is the standard gravitational parameter,
- r is the distance between the orbiting bodies.
- a is the length of the semi-major axis.

The velocity equation for a hyperbolic trajectory has either $+\frac{1}{a}$, or it is the same with the convention that in that case *a* is negative.

Energy

Under standard assumptions, specific orbital energy (ε) of elliptic orbit is negative and the orbital energy conservation equation (the Vis-viva equation) for this orbit can take the form:

$$\frac{v}{2} - \frac{\mu}{r} = -\frac{\mu}{2a} = \varepsilon < 0$$

where:

- v is orbital speed of orbiting body,
- r is distance of orbiting body from central body,
- a is length of semi-major axis,
- μ is standard gravitational parameter.

Conclusions:

- For a given semi-major axis the specific orbital energy is independent of the eccentricity.

Using the virial theorem we find:

- the time-average of the specific potential energy is equal to 2ε — the time-average of r^{-1} is a^{-1}

- the time-average of the specific kinetic energy is equal to -ε

Parabolic Orbits

If the eccentricity equals 1, then the orbit equation becomes:

$$r = \frac{h^2}{\mu}\frac{1}{1+\cos\theta}$$

where:

- r is radial distance of orbiting body from central body,
- h is specific angular momentum of the orbiting body,
- θ is a true anomaly of the orbiting body,
- μ is the standard gravitational parameter.

As the true anomaly θ approaches 180°, the denominator approaches zero, so that *r* tends towards infinity. Hence, the energy of the trajectory for which $e=1$ is zero, and is given by:

$$\varepsilon = \frac{v^2}{2} - \frac{\mu}{r} = 0$$

where:

- v is the orbital velocity of the orbiting body.

In other words, the speed anywhere on a parabolic path is:

$$v = \sqrt{\frac{2\mu}{r}}$$

Hyperbolic Orbits

If $e>1$, the orbit formula,

$$r = \frac{h^2}{\mu}\frac{1}{1+e\cos\theta}$$

describes the geometry of the hyperbolic orbit. The system consists of two symmetric curves. The orbiting body occupies one of them. The other one is its empty mathematical image. Clearly, the denominator of the equation above goes to zero when $\cos\theta = -1/e$. we denote this value of true anomaly

$$\theta_\infty = \cos^{-1}(-1/e)$$

since the radial distance approaches infinity as the true anomaly approaches θ_∞. θ_∞ is known as the *true anomaly of the asymptote.* Observe that θ_∞ lies between 90° and 180°. From the trig identity $\sin^2\theta+\cos^2\theta=1$ it follows that:

$$\sin\theta_\infty = (e^2-1)^{1/2}/e$$

Energy

Under standard assumptions, specific orbital energy (ε) of a hyperbolic trajectory is greater than zero and the orbital energy conservation equation for this kind of trajectory takes form:

$$\varepsilon = \frac{v^2}{2} - \frac{\mu}{r} = \frac{\mu}{-2a}$$

where:

- v is orbital velocity of orbiting body,
- r is radial distance of orbiting body from central body,
- a is the negative semi-major axis,
- μ is standard gravitational parameter.

Hyperbolic Excess Velocity

Under standard assumptions the body travelling along hyperbolic trajectory will attain in infinity an orbital velocity called hyperbolic excess velocity (v_∞) that can be computed as:

$$v_\infty = \sqrt{\frac{\mu}{-a}}$$

where:

- μ is standard gravitational parameter,
- a is the negative semi-major axis of orbit's hyperbola.

The hyperbolic excess velocity is related to the specific orbital energy or characteristic energy by

$$2\varepsilon = C_3 = v_\infty^2$$

Mathematical Techniques

Kepler's Equation

One approach to calculating orbits (mainly used historically) is to use Kepler's equation:

$$M = E - \varepsilon . \sin E.$$

where *M* is the mean anomaly, *E* is the eccentric anomaly, and is the eccentricity.

With Kepler's formula, finding the time-of-flight to reach an angle (true anomaly) of θ from periapsis is broken into two steps:

1. Compute the eccentric anomaly *E* from true anomaly θ
2. Compute the time-of-flight *t* from the eccentric anomaly *E*

Finding the eccentric anomaly at a given time (the inverse problem) is more difficult. Kepler's equation is transcendental in E, meaning it cannot be solved for E algebraically. Kepler's equation can be solved for E analytically by inversion.

Conic Orbits

For simple procedures, such as computing the delta-v for coplanar transfer ellipses, traditional approaches are fairly effective. Others, such as time-of-flight are far more complicated, especially for near-circular and hyperbolic orbits.

The Patched Conic Approximation

The Hohmann transfer orbit alone is a poor approximation for interplanetary trajectories because it neglects the planets' own gravity. Planetary gravity dominates the behaviour of the spacecraft in the vicinity of a planet and in most cases Hohmann severely overestimates delta-v, and produces highly inaccurate prescriptions for burn timings.

A relatively simple way to get a first-order approximation of delta-v is based on the 'Patched Conic Approximation' technique. One must choose the one dominant gravitating body in each region of space through which the trajectory will pass, and to model only that body's effects in that region. For instance, on a trajectory from the Earth to Mars, one would begin by considering only the Earth's gravity until the trajectory reaches a distance where the Earth's gravity no longer dominates that of the Sun. The spacecraft would be given escape velocity to send it on its way to interplanetary space. Next, one would consider only the Sun's gravity until the trajectory reaches the neighbourhood of Mars. During this stage, the transfer orbit model is appropriate. Finally, only Mars's gravity is considered during the final portion of the trajectory where Mars's gravity dominates the spacecraft's behaviour. The spacecraft would approach Mars on a hyperbolic orbit, and a final retrograde burn would slow the spacecraft enough to be captured by Mars.

The size of the "neighbourhoods" (or spheres of influence) vary with radius r_{SOI}:

$$r_{SOI} = a_p \left(\frac{m_p}{m_s} \right)^{2/5}$$

where a_p is the semimajor axis of the planet's orbit relative to the Sun; m_p and m_s are the masses of the planet and Sun, respectively.

This simplification is sufficient to compute rough estimates of fuel requirements, and rough time-of-flight estimates, but it is not generally accurate enough to guide a spacecraft to its destination. For that, numerical methods are required.

The Universal Variable Formulation

To address computational shortcomings of traditional approaches for solving the 2-body problem, the universal variable formulation was developed. It works equally well for the circular, elliptical, parabolic, and hyperbolic cases, the differential equations converging well when integrated for any orbit. It also generalises well to problems incorporating perturbation theory.

Perturbations

The universal variable formulation works well with the variation of parameters technique, except now, instead of the six Keplerian orbital elements, we use a different set of orbital elements: namely, the satellite's initial position and velocity vectors x_0 and v_0 at a given epoch $t = 0$. In a two-body simulation, these elements are sufficient to compute the satellite's position and velocity at any time in the future, using the universal variable formulation. Conversely, at any moment in the satellite's orbit, we can measure its position and velocity, and then use the universal variable approach to determine what its initial position and velocity *would have been* at the epoch. In perfect two-body motion, these orbital elements would be invariant (just like the Keplerian elements would be).

However, perturbations cause the orbital elements to change over time. Hence, we write the position element as $x_0(t)$ and the velocity element as $v_0(t)$, indicating that they vary with time. The technique to compute the effect of perturbations becomes one of finding expressions, either exact or approximate, for the functions $x_0(t)$ and $v_0(t)$.

Non-ideal Orbits

The following are some effects which make real orbits differ from the simple models based on a spherical earth. Most of them can be handled on short timescales (perhaps less than a few thousand orbits) by perturbation theory because they are small relative to the corresponding two-body effects.

- Equatorial bulges cause precession of the node and the perigee
- Tesseral harmonics of the gravity field introduce additional perturbations

- Lunar and solar gravity perturbations alter the orbits
- Atmospheric drag reduces the semi-major axis unless make-up thrust is used

Over very long timescales (perhaps millions of orbits), even small perturbations can dominate, and the behaviour can become chaotic. On the other hand, the various perturbations can be orchestrated by clever astrodynamicists to assist with orbit maintenance tasks, such as station-keeping, ground track maintenance or adjustment, or phasing of perigee to cover selected targets at low altitude.

6

Fluid Mechanics

Fluid mechanics is the study of fluids and the forces on them. (Fluids include liquids, gases, and plasmas.) Fluid mechanics can be divided into fluid statics, the study of fluids at rest; fluid kinematics, the study of fluids in motion; and fluid dynamics, the study of the effect of forces on fluid motion. It is a branch of continuum mechanics, a subject which models matter without using the information that it is made out of atoms, that is, it models matter from a macroscopic viewpoint rather than from a microscopic viewpoint. Fluid mechanics, especially fluid dynamics, is an active field of research with many unsolved or partly solved problems. Fluid mechanics can be mathematically complex. Sometimes it can best be solved by numerical methods, typically using computers. A modern discipline, called computational fluid dynamics (CFD), is devoted to this approach to solving fluid mechanics problems. Also taking advantage of the highly visual nature of fluid flow is particle image velocimetry, an experimental method for visualising and analysing fluid flow.

Brief History

The study of fluid mechanics goes back at least to the days of ancient Greece, when Archimedes investigated fluid statics and buoyancy and formulated his famous law known now as the Archimedes' Principle. Rapid advancement in fluid mechanics began with Leonardo da Vinci (observation and experiment), Evangelista Torricelli (barometer), Isaac Newton (viscosity) and Blaise Pascal (hydrostatics), and was continued by Daniel Bernoulli with the introduction of mathematical fluid dynamics in *Hydrodynamica* (1738). Inviscid flow was further analysed by various mathematicians (Leonhard Euler, d'Alembert, Lagrange, Laplace, Poisson) and viscous flow was explored by a multitude of engineers including Poiseuille and Gotthilf Heinrich Ludwig Hagen. Further mathematical justification was provided by Claude-Louis Navier and George Gabriel Stokes in the Navier–Stokes equations, and boundary layers were investigated (Ludwig Prandtl, Theodore von Kármán), while various scientists (Osborne Reynolds, Andrey Kolmogorov, Geoffrey Ingram Taylor) advanced the understanding of fluid viscosity and turbulence.

Assumptions

Like any mathematical model of the real world, fluid mechanics makes some basic assumptions about the materials being studied. These assumptions are turned into equations that must be satisfied if the assumptions are to be held true. For example, consider an incompressible fluid in three dimensions. The assumption that mass is conserved means that for any fixed closed surface (such as a sphere) the rate of mass passing from *outside* to *inside* the surface must be the same as rate of mass passing the other way. (Alternatively, the mass *inside* remains constant, as does the mass *outside*). This can be turned into an integral equation over the surface.

Fluid mechanics assumes that every fluid obeys the following:

- Conservation of mass
- Conservation of energy
- Conservation of momentum.

Further, it is often useful (at subsonic conditions) to assume a fluid is incompressible – that is, the density of the fluid does not change.

Similarly, it can sometimes be assumed that the viscosity of the fluid is zero (the fluid is *inviscid*). Gases can often be assumed to be inviscid. If a fluid is viscous, and its flow contained in some way (e.g. in a pipe), then the flow at the boundary must have zero velocity. For a viscous fluid, if the boundary is not porous, the shear forces between the fluid and the boundary results also in a zero velocity for the fluid at the boundary. This is called the no-slip condition. For a porous media otherwise, in the frontier of the containing vessel, the slip condition is not zero velocity, and the fluid has a discontinuous velocity field between the free fluid and the fluid in the porous media (this is related to the Beavers and Joseph condition).

The Continuum Hypothesis

Fluids are composed of molecules that collide with one another and solid objects. The continuum assumption, however, considers fluids to be continuous. That is, properties such as density, pressure, temperature, and velocity are taken to be well-defined at "infinitely" small points, defining a REV (Reference Element of Volume), at the geometric order of the distance between two adjacent molecules of fluid. Properties are assumed to vary continuously from one point to another, and are averaged values in the REV. The fact that the fluid is made up of discrete molecules is ignored.

The continuum hypothesis is basically an approximation, in the same way planets are approximated by point particles when dealing with celestial mechanics, and therefore results in approximate solutions. Consequently, assumption of the continuum hypothesis can lead to results which are not of desired accuracy. That said, under the right circumstances, the continuum hypothesis produces extremely accurate results.

Those problems for which the continuum hypothesis does not allow solutions of desired accuracy are solved using statistical mechanics. To determine whether or not to use conventional fluid dynamics or statistical mechanics, the Knudsen number is evaluated for the problem. The Knudsen number is defined as the ratio of the molecular mean free path length to a certain representative physical length scale. This length scale could be, for example, the radius of a body in a fluid. (More simply, the Knudsen number is how many times its own diameter a particle will travel on average before hitting another particle). Problems with Knudsen numbers at or above unity are best evaluated using statistical mechanics for reliable solutions.

Navier–Stokes Equations

The Navier–Stokes equations (named after Claude-Louis Navier and George Gabriel Stokes) are the set of equations that describe the motion of fluid substances such as liquids and gases. These equations state that changes in momentum (force) of fluid particles depend only on the external pressure and internal viscous forces (similar to friction) acting on the fluid. Thus, the Navier–Stokes equations describe the balance of forces acting at any given region of the fluid.

The Navier–Stokes equations are differential equations which describe the motion of a fluid. Such equations establish relations among the rates of change of the variables of interest. For example, the Navier–Stokes equations for an ideal fluid with zero viscosity states that acceleration (the rate of change of velocity) is proportional to the derivative of internal pressure.

This means that solutions of the Navier–Stokes equations for a given physical problem must be sought with the help of calculus. In practical terms only the simplest cases can be solved exactly in this way. These cases generally involve non-turbulent, steady flow (flow does not change with time) in which the Reynolds number is small.

For more complex situations, such as global weather systems like El Niño or lift in a wing, solutions of the Navier–Stokes equations can currently only be found with the help of computers. This is a field of sciences by its own called computational fluid dynamics.

Newtonian versus Non-Newtonian Fluids

A Newtonian fluid (named after Isaac Newton) is defined to be a fluid whose shear stress is linearly proportional to the velocity gradient in the direction perpendicular to the plane of shear. This definition means regardless of the forces acting on a fluid, it *continues to flow*. For example, water is a Newtonian fluid, because it continues to display fluid properties no matter how much it is stirred or mixed. A slightly less rigorous definition is that the drag of a small object being moved slowly through the fluid is proportional to the force applied to the object. (Compare friction). Important fluids, like water as well as most gases, behave — to good approximation — as a Newtonian fluid under normal conditions on Earth.

By contrast, stirring a non-Newtonian fluid can leave a "hole" behind. This will gradually fill up over time – this behaviour is seen in materials such as pudding, oobleck, or sand (although sand isn't strictly a fluid). Alternatively, stirring a non-Newtonian fluid can cause the viscosity to decrease, so the fluid appears "thinner" (this is seen in non-drip paints). There are many types of non-Newtonian fluids, as they are defined to be something that fails to obey a particular property — for example, most fluids with long molecular chains can react in a non-Newtonian manner.

Equations for a Newtonian Fluid

The constant of proportionality between the shear stress and the velocity gradient is known as the viscosity. A simple equation to describe Newtonian fluid behaviour is;

$$\tau = -\mu \frac{dv}{dy}$$

where

τ is the shear stress exerted by the fluid ("drag")

μ is the fluid viscosity – a constant of proportionality $\frac{dv}{dy}$ is the velocity gradient perpendicular to the direction of shear.

For a Newtonian fluid, the viscosity, by definition, depends only on temperature and pressure, not on the forces acting upon it. If the fluid is incompressible and viscosity is constant across the fluid, the equation governing the shear stress (in Cartesian coordinates) is

$$\tau_{ij} = -\mu\left(\frac{\partial v_i}{\partial x_j} + \frac{\partial u_j}{\partial x_i}\right)$$

where

t_{ij} is the shear stress on the ith face of a fluid element in the jth direction

v_i is the velocity in the ith direction

x_j is the jth direction coordinate.

If a fluid does not obey this relation, it is termed a non-Newtonian fluid, of which there are several types.

Among fluids, two rough broad divisions can be made: ideal and non-ideal fluids. An ideal fluid really does not exist, but in some calculations, the assumption is justifiable. An Ideal fluid is non viscous-offers no resistance whatsoever to a shearing force.

One can group real fluids into Newtonian and non-Newtonian. Newtonian fluids agree with Newton's law of viscosity. Non-Newtonian fluids can be either plastic, bingham plastic, pseudoplastic, dilatant, thixotropic, rheopectic, viscoelatic.

Soil Mechanics

Soil mechanics is a branch of engineering mechanics that describes the behaviour of soils. It differs from fluid mechanics and solid mechanics in the sense that soils consist of a heterogeneous mixture of fluids (usually air and water) and particles (usually clay, silt, sand, and gravel) but soil may also contain organic solids, liquids, and gasses and other matter. Along with rock mechanics, soil mechanics provides the theoretical basis for analysis in geotechnical engineering, a subdiscipline of Civil engineering. Soil mechanics is used to analyse the deformations of and flow of fluids within natural and man-made structures that are supported on or made of soil, or structures that are buried in soils. Examples applications are building and bridge foundations, retaining walls, dams, and buried pipeline systems.

Principles of soil mechanics are also used in related disciplines such as geophysical engineering, engineering geology, coastal engineering, agricultural engineering, hydrology and soil physics.

The genesis and composition of soil, the distinction between *pore water pressure* and inter-granular *effective stress*, capillary action of fluids in the pore spaces, *soil classification*, *seepage* and *permeability*, time dependent change of volume due to squeesing water out of tiny pore spaces, also known as *consolidation*, *shear strength* and stiffness of soils. The shear strength of soils is primarily derived from friction between the particles and interlocking, which are very sensitive to the effective stress. Some examples of geotechnical engineering applications of the principles of soil mechanics such as slope stability, lateral earth pressure on retaining walls, and bearing capacity of foundations.

Genesis and Composition of Soils

Genesis: The primary mechanism of soil creation is the weathering of rock. All rock types (igneous rock, metamorphic rock and sedimentary rock) may be broken down into small particles to create soil. Weathering mechanisms are physical weathering, chemical weathering, and biological weathering Human activities such as excavation, blasting, and waste disposal, may also create soil. Over geologic time, deeply buried soils may be altered by pressure and temperature to become metamorphic or sedimentary rock, and if melted and solidified again, they would complete the geologic cycle by becoming igneous rock.

Physical weathering includes temperature effects, freeze and thaw of water in cracks, rain, wind, impact and other mechanisms. Chemical weathering includes dissolution of matter composing a rock and precipitation in the form of another mineral. Clay minerals, for example can be formed by weathering of feldspar, which is the most common mineral present in igneous rock. The most common mineral constituent of silt and sand is quartz, also called silica, which has the chemical name silicon dioxide. The reason that feldspar is most common in rocks but silicon is more prevalent in soils is that feldspar is much more soluble than silica.

Silt, Sand, and Gravel are basically little pieces of broken rocks.

According to the Unified Soil Classification System, silt particle sizes are in the range of 0.002 mm to 0.075 mm and sand particles have sizes in the range of 0.075 mm to 4.75 mm.

Gravel particles are broken pieces of rock in the size range 4.75 mm to 100 mm.

Particles larger than gravel are called cobbles and boulders.

Transport

Soil deposits are affected by the mechanism of transport and deposition to their location. Soils that are not transported are called residual soils — they exist at the same location as the rock from which they were generated. Decomposed granite is a common example of a residual soil. The common mechanisms of transport are the actions of gravity, ice, water, and wind. Wind blown soils include dune sands and loess. Water carries particles of different size depending on the speed of the water, thus soils transported by water are graded according to their size. Silt and clay may settle out in a lake, and gravel and sand collect at the bottom of a river bed. Wind blown soil deposits (aeolian soils) also tend to be sorted according to their grain size. Erosion at the base of glaciers is powerful enough to pick up large rocks and boulders as well as soil; soils dropped by melting ice can be a well graded mixture of widely varying particle sizes. Gravity on its own may also carry particles down from the top of a mountain to make a pile of soil and boulders at the base; soil deposits transported by gravity are called colluvium.

The mechanism of transport also has a major effect on the particle shape. For example, low velocity grinding in a river bed will produce rounded particles. Freshly fractured colluvium particles often have a very angular shape.

Soil Composition

Soil Mineralogy: Silts, sands and gravels are classified by their size, and hence they may consist of a variety of minerals. Owing to the stability of quartz compared to other rock minerals, quartz is the most common constituent of sand and silt . Mica, and feldspar are other common minerals present in sands and silts. The mineral constituents of gravel may be more similar to that of the parent rock.

The common clay minerals are montmorillonite or smectite, illite, and kaolinite or kaolin. These minerals tend to form in sheet or plate like structures, with length typically ranging between $10^{-7}m$ and $4x10^{-6}m$ and thickness typically ranging between $10^{-9}m$ and $2x10^{-6}m$, and they have a relatively large specific surface area. The specific surface area (SSA) is defined as the ratio of the surface area of particles to the mass of the particles. Clay minerals typically have specific surface areas in the range of 10 to 1,000 square meters per gram of solid. Due to the large surface area available for chemical, electrostatic, and van der Waals interaction, the mechanical behaviour of clay minerals is very sensitive to the amount of pore fluid available and the type and amount of dissolved ions in the pore fluid.

The minerals of soils are predominantly formed by atoms of oxygen, silicon, hydrogen, and aluminium, organised in various crystalline forms. These elements along with calcium, sodium, potassium, magnesium, and carbon constitute over 99 per cent of the solid mass of soils.

Grain Size Distribution

Soils consist of a mixture of particles of different size, shape and mineralogy. Because the size of the particles obviously has a significant effect on the soil behaviour, the grain size and grain size distribution are used to classify soils. The grain size distribution describes the relative proportions of particles of various sizes. The grain size is often visualised in a cumulative distribution graph which, for example, plots the percentage of particles finer than a given size as a function of size. The median grain size, D_{50}, is the size for which 50% of the particle mass consists of finer particles. Soil behaviour, especially the hydraulic conductivity, tends to be dominated by the smaller particles, hence, the term "effective size", denoted by D_{10}, is defined as the size for which 10% of the particle mass consists of finer particles.

Sands and gravels that possess a wide range of particle sizes with a smooth distribution of particle sizes are called *well graded* soils. If the soil particles in a sample are predominantly in a relatively narrow range of sizes, the soil are called *uniformly graded* soils. If there are distinct gaps in the gradation curve, e.g., a mixture of gravel and fine sand, with no coarse sand, the soils may be called *gap graded. Uniformly graded* and *gap graded* soils are both considered to be *poorly graded.* There are many methods for measuring particle size distribution. The two traditional methods used in geotechnical engineering are sieve analysis and hydrometer analysis.

Sieve Analysis

The size distribution of gravel and sand particles are typically measured using sieve analysis. The formal procedure is described in ASTM D6913-04(2009). A stack of sieves with accurately dimensioned holes between a mesh of wires is used to separate the particles into size bins. A known volume of dried soil, with clods broken down to individual particles, is put into the top of a stack of sieves arranged from coarse to fine. The stack of sieves is shaken for a standard period of time so that the particles are sorted into size bins. This method works reasonably well for particles in the sand and gravel size range. Fine particles tend to stick to each other, and hence the sieving process is not an effective method. If there are a lot of fines (silt and clay) present in the soil it may be necessary to run water through the sieves to wash the coarse particles and clods through.

A variety of sieve sizes are available. The boundary between sand and silt is arbitrary. According to the Unified Soil Classification System, a #4 sieve (4 openings per inch) having 4.75mm opening size separates sand from gravel and a #200 sieve with an 0.075 mm opening separates sand from silt and clay. According to the British standard, 0.063 mm is the boundary between sand and silt, and 2 mm is the boundary between sand and gravel.

Hydrometer Analysis

The classification of fine-grained soils, i.e., soils that are finer than sand, is determined primarily by their Atterberg limits, not by their grain size. If it is important to determine the grain size distribution of fine-grained soils, the hydrometer test may be performed. In the hydrometer tests, the soil particles are mixed with water and shaken to produce a dilute suspension in a glass cylinder, and then the cylinder is left to sit. A hydrometer is used to measure the density of the suspension as a function of time. Clay particles may take several hours to settle past the depth of measurement of the hydrometer. Sand particles may take less than a second. Stoke's law provides the theoretical basis to calculate the relationship between sedimentation velocity and particle size. ASTM provides the detailed procedures for performing the Hydrometer test.

Clay particles can be sufficiently small that they never settle because they are kept in suspension by Brownian motion, in which case they may be classified as colloids.

Effective Stress and Capillarity: Hydrostatic Conditions

To understand the mechanics of soils it is necessary to understand how normal stresses and shear stresses are shared by the different phases. Neither gas nor liquid provide significant resistance to shear stress. The shear resistance of soil is provided by friction and interlocking of the particles. The friction depends on the intergranular contact stresses between solid particles. The normal stresses, on the other hand, are shared by the fluid and the particles. Although the pore air is relatively compressible, and hence takes little normal stress in most geotechnical engineering problems, liquid water is relatively incompressible and if the voids are saturated with water, the pore water must be squeezed out in order to pack the particles closer together.

The principle of effective stress, introduced by Karl Terzaghi, states that the effective stress σ' (i.e., the average intergranular stress between solid particles) may be calculated by a simple subtraction of the pore pressure from the total stress:

$$\sigma' = \sigma - u$$

where σ is the total stress and u is the pore pressure. It is not practical to measure σ' directly, so in practice the vertical effective stress is calculated from the pore pressure and vertical total stress. The distinction between the terms pressure and stress is also important. By definition, pressure at a point is equal in all directions but stresses at a point can be different in different directions. In soil mechanics, compressive stresses and pressures are considered to be positive and tensile stresses are considered to be negative, which is different from the solid mechanics sign convention for stress.

Total Stress

For level ground conditions, the total vertical stress at a point, σ_v, on average, is the weight of everything above that point per unit area. The vertical stress beneath a uniform surface layer with density ρ, and thickness H is for example:

$$\sigma_v = \rho g H = \gamma H$$

where g is the acceleration due to gravity, and γ is the unit weight of the overlying layer. If there are multiple layers of soil or water above the point of interest, the vertical stress may be calculated by summing the product of the unit weight and thickness of all of the overlying layers. Total stress increases with increasing depth in proportion to the density of the overlying soil. It is not possible to calculate the horizontal total stress in this way. Lateral earth pressures are addressed elsewhere.

Hydrostatic Conditions

If there is no pore water flow occurring in the soil, the pore water pressures will be hydrostatic. The water table is located at the depth where the water pressure is equal to the atmospheric pressure. For hydrostatic conditions, the water pressure increases linearly with depth below the water table:

$$u = \rho_w g z_w$$

where ρ_w is the density of water, and z_w is the depth below the water table.

Capillary Action

Due to surface tension water will rise up in a small capillary tube above a free surface of water. Likewise, water will rise up above the water table into the small pore spaces around the soil particles. In fact the soil may be completely saturated for some distance above the water table. Above the height of capillary saturation, the soil may be wet but the water content will decrease with elevation. If the water in the capillary zone is not moving, the water pressure obeys the equation of hydrostatic equilibrium, $u = \rho_w g z_w$, but note that z_w, is negative above the water table. Hence, hydrostatic water pressures are negative above the water table. The thickness of the zone of capillary saturation depends on the pore size, but typically, the heights vary between a centimetre or so for coarse sand to tens of meters for a silt or clay.

The surface tension of water explains why the water does not drain out of a wet sand castle or a moist ball of clay. Negative water pressures make the water stick to the particles and pull the particles to each other, friction at the particle contacts make a sand castle stable. But as soon as a wet sand castle is submerged below a free water surface, the negative pressures are lost and the castle collapses. Considering the effective stress equation, $\sigma' = \sigma - u$,, if the water pressure is negative, the effective stress may be positive, even on a free surface (a surface where the total normal stress is zero). The negative pore pressure pulls the particles together and causes compressive particle to particle contact forces.

Negative pore pressures in clayey soil can be much more powerful than those in sand. Negative pore pressures explain why clay soils shrink when they dry and swell as they are wetted. The swelling and shrinkage can cause major distress, especially to light structures and roads.

Soil Classification

Geotechnical engineers classify the soil particle types by performing tests on disturbed (dried, passed through sieves, and remodelled) samples of the soil. This provides information about the characteristics of the soil grains themselves. It should be noted that classification of the types of grains present in a soil does not account for important effects of the *structure* or *fabric* of the soil, terms that describe compactness of the particles and patterns in the arrangement of particles in a load carrying framework as well as the pore size and pore fluid distributions.

Classification of Soil Grains

In the US and other countries, the Unified Soil Classification System (USCS)is often used for soil classification. Other classification systems include the British Standard BS5390 and the AASHTO soil classification system.

Classification of Sands and Gravels

In the USCS, gravels (given the symbol *G*) and sands (given the symbol *S*) are classified according to their grain size distribution. For the USCS, gravels may be given the classification symbol *GW* (well-graded gravel), *GP* (poorly graded gravel), *GM* (gravel with a large amount of silt), or *GC* (gravel with a large amount of clay). Likewise sands may be classified as being *SW*, *SP*, *SM* or *SC*. Sands and gravels with a small but non-negligible amount of fines (5 % - 12 %) may be given a dual classification such as *SW-SC*.

Atterberg Limits

Clays and Silts, often called 'fine-grained soils', are classified according to their Atterberg limits; the most commonly used Atterberg limits are the Liquid limit (denoted by *LL* or w_l), Plastic Limit (denoted by *PL* or w_p), and Shrinkage limit (denoted by *SL*). The shrinkage limit corresponds to a water content below which the soil will not shrink as it dries.

The liquid limit and plastic limit are arbitrary limits determined by tradition and convention. The liquid limit is determined by measuring the water content for which a groove closes after 25 blows in a standard test. Alternatively, a fall cone test apparatus may be use to measure the liquid Limit.

The undrained shear strength of remodelled soil at the liquid limit is approximately 2 kPa. The plastic limit is the water content below which it is not possible to roll by hand the soil into 3 mm diameter cylinders. The soil cracks or breaks up as it is rolled down to this diameter. Remodelled soil at the plastic limit is quite stiff, having an undrained shear strength of the order of about 200 kPa.

The Plasticity index of a particular soil specimen is defined as the difference between the Liquid limit and the Plastic limit of the specimen; it is an indicator of how much water the soil particles in the specimen can absorb. The plasticity index is the difference in water contents between states when the soil is relatively soft and the soil is relatively brittle when molded by hand.

Classification of Silts and Clays

According to the Unified Soil Classification System (USCS), silts and clays are classified by plotting the values of their plasticity index and liquid limit on a plasticity chart. The A-Line on the chart separates clays (given the USCS symbol *C*) from silts (given the symbol *M*). LL=50% separates high plasticity soils (given the modifier symbol *H*) from low plasticity soils (given the modifier symbol *L*). A soil that plots above the A-line and has LL>50% would, for example, be classified as *CH*. Other possible classifications of silts and clays are *ML*, *CL* and *MH*. If the Atterberg limits plot in the "hatched" region on the graph near the origin, the soils are given the dual classification 'CL-ML'.

Indices Related to Soil Strength

Liquidity Index

The effects of the water content on the strength of saturated remodelled soils can be quantified by the use of the *liquidity index*, *LI*:

$$\mathrm{LI} = \frac{\mathrm{w} - \mathrm{PL}}{\mathrm{LL} - \mathrm{PL}}$$

When the LI is 1, remodelled soil is at the liquid limit and it has an undrained shear strength of about 2 kPa. When the soil is at the plastic limit, the LI is 0 and the undrained shear strength is about 200 kPa.

Relative Density

The density of sands (cohesionless soils) is often characterised by the relative density, D_r

$$\mathrm{Dr} = \frac{e_{max} - e}{e_{max} - e_{min}} 100\%$$

where: e_{max} is the "maximum void ratio" corresponding to a very loose state, e_{min} is the "minimum void ratio" corresponding to a very dense state and e is the in situ void ratio. Methods used to calculate relative density are defined in ASTM D4254-00(2006).

Thus if D_r = 100% the sand or gravel is very dense, and if D_r = 0% the soil is extremely loose and unstable.

Seepage: Steady State Flow of Water

If fluid pressures in a soil deposit are uniformly increasing with depth according to $u = \rho_w g z_w$ then hydrostatic conditions will prevail and the fluids will not be flowing through the soil. z_w is the depth. However, if the water table is sloping or there is a perched water table as indicated in the accompanying sketch, then seepage will occur. For steady state seepage, the seepage velocities are not varying with time. If the water tables are changing levels with time, or if the soil is in the process of consolidation, then steady state conditions do not apply.

Darcy's Law

Darcy's law states that the volume of flow of the pore fluid through a porous medium per unit time is proportional to the rate of change of excess fluid pressure with distance. The constant of proportionality includes the viscosity of the fluid and the intrinsic permeability of the soil. For the simple case of a horizontal tube filled with soil

$$Q = \frac{-KA}{\mu} \frac{(u_b - u_a)}{L}$$

The total discharge, Q (units of volume per time, e.g., ft^3/s or m^3/s) is proportional to the intrinsic permeability, K, the cross sectional area, A, and rate of pore pressure change with distance, $\frac{(u_b - u_a)}{L}$, and inversely proportional to the dynamic viscosity of the fluid, μ. The negative sign is needed because fluids flow from high pressure to low pressure. So if the change in pressure is negative (in the x-direction) then the flow will be positive (in the x-direction). The above equation works well for a horizontal tube, but if the tube was inclined so that point b was a different elevation than point a, the equation would not work. The effect of elevation is accounted for by replacing the pore pressure by *excess pore pressure*, u_e defined as:

$$u_e = u - \rho_w g z$$

where z is the depth measured from an arbitrary elevation reference (datum). Replacing u by u_e we obtain a more general equation for flow:

$$Q = \frac{-KA}{\mu} \frac{(u_{e,b} - u_{e,a})}{L}$$

Dividing both sides of the equation by A, and expressing the rate of change of excess pore pressure as a derivative, we obtain a more general equation for the apparent velocity in the x-direction:

$$v_x = \frac{-K}{\mu}\frac{du_e}{dx}$$

where v_x has units of velocity and is called the *Darcy velocity*, or *discharge velocity*. The *seepage velocity* v_{sx} = (average velocity of fluid molecules in the pores) is related to the Darcy velocity, and the porosity, n

$$v_{sx} = \frac{v_x}{n}$$

Civil engineers predominantly work on problems that involve water and predominantly work on problems on earth (in earths gravity). For this class of problems, civil engineers will often write Darcy's law in a much simpler form :

$$v = ki$$

where k is called *permeability*, and is defined as $k = \frac{K\rho_w g}{\mu_w}$, and i is called the *hydraulic gradient*. The hydraulic gradient is the rate of change of total head with distance. The total head, h at a point is defined as the height (measured relative to the datum) to which water would raise in a piezometer at that point. The total head is related to the excess water pressure by:

$$u_e = \rho_w gh + Constant$$

and the *Constant* is zero if the datum for head measurement is chosen at the same elevation as the origin for the depth, z used to calculate u_e.

Typical Values of Permeability

Values of the permeability, k, can vary by many orders of magnitude depending on the soil type. Clays may have permeability as small as about $10^{-12}\frac{m}{s}$, gravels may have permeability up to about $10^{-1}\frac{m}{s}$. Layering and heterogeneity and disturbance during the sampling and testing process make the accurate measurement of soil permeability a very difficult problem.

Flow Nets

Darcy's Law applies in one, two or three dimensions. In two or three dimensions, steady state seepage is described by Laplace's equation. Computer programmes are available to solve this equation. But traditionally two-dimensional seepage problems were solved using and a graphical procedure known called flow nets. One set of lines in the flow net are in the direction of the water flow (flow lines), and the other set of lines are in the direction of constant total head (equipotential lines). Flow nets may be used for example to estimate the quantity of seepage under dams and sheet piling.

Seepage Forces and Erosion

When the seepage velocity is great enough, erosion can occur because of the frictional drag exerted on the soil particles. Vertically upwards seepage is a source of danger on the downstream side of sheet piling and beneath the toe of a dam or levee. Erosion of the soil, known as "piping", can lead to failure of the structure and to sinkhole formation. Seeping water removes soil, starting from the exit point of the seepage, and erosion advances upgradient. The term sand boil is used to describe the appearance of the discharging end of an active soil pipe.

Seepage Pressures

Seepage in an upward direction reduces the effective stress within the soil. When the water pressure at a point in the soil is equal to the total vertical stress at that point, the effective stress is zero and the soil has no frictional resistance to deformation. For a surface layer, the vertical effective stress becomes zero within the layer when the upward hydraulic gradient is equal to the critical gradient. At zero effective stress soil has very little strength and layers of relatively impermeable soil may heave up due to the underlying water pressures. The loss in strength due to upward seepage is a common contributor to levee failures. The condition of zero effective stress associated with upward seepage is also called liquefaction, quicksand, or a boiling condition. Quicksand was so named because the soil particles move around and appear to be 'alive' (the biblical meaning of 'quick' - as opposed to 'dead'). (Note that it is not possible to be 'sucked down' into quicksand. On the contrary, you would float with about half your body out of the water.)

Consolidation: Transient Flow of Water

Consolidation is a process by which soils decrease in volume. It occurs when stress is applied to a soil that causes the soil particles to pack together more tightly, therefore reducing volume. When this occurs in a soil that is saturated with water, water will be squeezed out of the soil. The time required to squeeze the water out of a thick deposit of clayey soil layer might be years. For a layer of sand, the water may be squeezed out in a matter of seconds. A building foundation or construction of a new embankment will cause the soil below to consolidate and this will cause settlement which in turn may cause distress to the building or embankment. Karl Terzaghi developed the theory of consolidation which enables prediction of the amount of settlement and the time required for the settlement to occur. Soils are tested with an oedometer test to determine their compression index and coefficient of consolidation.

When stress is removed from a consolidated soil, the soil will rebound, drawing water back into the pores and regaining some of the volume it had lost in the consolidation process. If the stress is reapplied, the soil will re-consolidate again along a recompression curve, defined by the recompression index. Soil that has been consolidated to a large pressure and has been subsequently unloaded is considered to be *overconsolidated.* The maximum past vertical effective stress is termed the *preconsolidation stress.* A soil which is currently experiencing the maximum past vertical effective stress is said to be *normally consolidated.* The *overconsolidation ratio,* (OCR) is the ratio of the maximum past vertical effective stress to the current vertical effective stress. The OCR is significant for two reasons: firstly, because the compressibility of normally consolidated soil is significantly larger than that for overconsolidated soil, and secondly, the shear behaviour and dilatancy of clayey soil are related to the OCR through critical state soil mechanics; highly overconsolidated clayey soils are dilatant, while normally consolidated soils tend to be contractive.

Shear Behaviour: Stiffness and Strength

The shear strength and stiffness of soil determines whether or not soil will be stable or how much it will deform. Knowledge of the strength is necessary to determine if a slope will be stable, if a building or bridge might settle too far into the ground, and the limiting pressures on a retaining wall. It is important to distinguish between failure of a soil element and the failure of a geotechnical structure (e.g., a building foundation, slope or retaining wall); some soil elements may reach their peak strength prior to failure of the structure. Different criteria can be used to define the "shear strength" and the "yield point" for a soil element from a stress-strain curve. One may define the peak shear strength as the peak of a stress strain curve, or the shear strength at critical state as the value after large strains when the shear resistance levels off. If the stress-strain curve does not stabilise before the end of shear strength test, the "strength" is sometimes considered to be the shear resistance at 15% to 20% strain. The shear strength of soil depends on many factors including the effective stress and the void ratio.

The shear stiffness is important, for example, for evaluation of the magnitude of deformations of foundations and slopes prior to failure and because it is related to the shear wave velocity. The slope of the initial, nearly linear, portion of a plot of shear stress as a function of shear strain is called the shear modulus

Friction, Interlocking and Dilation

Soil is an assemblage of particles that have little to no cementation while rock (such as sandstone) may consist of an assembly of particles that are strongly cemented together by chemical bonds. The shear strength of soil is primarily due to interparticle friction and therefore, the shear resistance on a plane is approximately proportional to the effective normal stress on that plane. But soil also derives significant shear resistance from interlocking of grains. If the grains are densely packed, the grains tend to spread apart from each other as they are subject to shear strain. The expansion of the particle matrix due to shearing was called dilatancy by Osborne Reynolds. If one considers the energy required to shear an assembly of particles there is energy input by the shear force, T, moving a distance, x and there is also energy input by the normal force, N, as the sample expands a distance, y. Due to the extra energy required for the particles to dilate against the confining pressures, dilatant soils have a greater peak strength than contractive soils. Furthermore, as dilative soil grains dilate, they become looser (their void ratio increases), and their rate of dilation decreases until they reach a critical void ratio. Contractive soils become denser as they shear, and their rate of contraction decreases until they reach a critical void ratio.

The tendency for a soil to dilate or contract depends primarily on the confining pressure and the void ratio of the soil. The rate of dilation is high if the confining pressure is small and the void ratio is small. The rate of contraction is high if the confining pressure is large and the void ratio is large. As a first approximation, the regions of contraction and dilation are separated by the critical state line.

Structure, Fabric, and Chemistry

In addition to the friction and interlocking (dilatancy) components of strength, the structure and fabric also play a significant role in the soil behaviour. The structure and fabric include factors such as the spacing and arrangement of the solid particles or the amount and spatial distribution of pore water; in some cases cementitious material accumulates at particle-particle contacts. Mechanical behaviour of soil is affected by the density of the particles and their structure or arrangement of the particles as well as the amount and spatial distribution of fluids present (e.g., water and air voids). Other factors include the electrical charge of the particles, chemistry of pore water, chemical bonds (i.e. cementation -particles connected through a solid substance such as recrystallised calcium carbonate)

Drained and Undrained Shear

The presence of nearly incompressible fluids such as water in the pore spaces affects the ability for the pores to dilate or contract.

If the pores are saturated with water, water must be sucked into the dilating pore spaces to fill the expanding pores (this phenomena is visible at the beach when apparently dry spots form around feet that press into the wet sand).

Similarly, for contractive soil, water must be squeezed out of the pore spaces to allow contraction to take place.

Dilation of the voids causes negative water pressures that draw fluid into the pores, and contraction of the voids causes positive pore pressures to push the water out of the pores. If the rate of shearing is very large compared to the rate that water can be sucked into or squeezed out of the dilating or contracting pore spaces, then the shearing is called *undrained shear*, if the shearing is slow enough that the water pressures are negligible, the shearing is called *drained shear*. During undrained shear, the water pressure u changes depending on volume change tendencies. From the effective stress equation, the change in u directly effects the effective stress by the equation:

$$\sigma' = \sigma - u$$

and the strength is very sensitive to the effective stress. It follows then that the undrained shear strength of a soil may be smaller or larger than the drained shear strength depending upon whether the soil is contractive or dilative.

Shear Tests

Strength parameters can be measured in the laboratory using direct shear test, triaxial shear test, simple shear test, fall cone test and vane shear test; there are numerous other devices and variations on these devices used in practice today. Tests conducted to characterise the strength and stiffness of the soils in the ground include the Cone penetration test and the Standard penetration test.

Other Factors

The stress-strain relationship of soils, and therefore the shearing strength, is affected by :

1. *soil composition* (basic soil material): mineralogy, grain size and grain size distribution, shape of particles, pore fluid type and content, ions on grain and in pore fluid.
2. *state* (initial): Define by the initial void ratio, effective normal stress and shear stress (stress history). State can be describe by terms such as: loose, dense, overconsolidated, normally consolidated, stiff, soft, contractive, dilative, etc.
3. *structure*: Refers to the arrangement of particles within the soil mass; the manner in which the particles are packed or distributed. Features such as layers, joints, fissures, slickensides, voids, pockets, cementation, etc., are part of the structure. Structure of soils is described by terms such as: undisturbed, disturbed, remodelled, compacted, cemented; flocculent, honey-combed, single-grained; flocculated, deflocculated; stratified, layered, laminated; isotropic and anisotropic.
4. *Loading conditions*: Effective stress path -drained, undrained, and type of loading -magnitude, rate (static, dynamic), and time history (monotonic, cyclic).

Applications

Lateral Earth Pressure: Lateral earth stress theory is used to estimate the amount of stress soil can exert perpendicular to gravity. This is the stress exerted on retaining walls. A lateral earth stress coefficient, K, is defined as the ratio of lateral (horizontal) stress to vertical stress for cohesionless soils ($K=\sigma_h/\sigma_v$). There are three coefficients: at-rest, active, and passive. At-rest stress is the lateral stress in the ground before any disturbance takes place. The active stress state is reached when a wall moves away from the soil under the influence of lateral stress, and results from shear failure due to reduction of lateral stress. The passive stress state is reached when a wall is pushed into the soil far enough to cause shear failure within the mass due to increase of lateral stress. There are many theories for estimating lateral earth stress; some are empirically based, and some are analytically derived.

Bearing Capacity: The bearing capacity of soil is the average contact stress between a foundation and the soil which will cause shear failure in the soil. Allowable bearing stress is the bearing capacity divided by a factor of safety. Sometimes, on soft soil sites, large settlements may occur under loaded foundations without actual shear failure occurring; in such cases, the allowable bearing stress is determined with regard to the maximum allowable settlement.

Slope Stability

The field of slope stability encompasses the analysis of static and dynamic stability of slopes of earth and rock-fill dams, slopes of other types of embankments, excavated slopes, and natural slopes in soil and soft rock.

As seen to the right, earthen slopes can develop a cut-spherical weakness zone. The probability of this happening can be calculated in advance using a simple 2-D circular analysis package... A primary difficulty with analysis is locating the most-probable slip plane for any given situation. Many landslides have been analysed only after the fact.

Continuum Mechanics

Continuum mechanics is a branch of mechanics that deals with the analysis of the kinematics and the mechanical behaviour of materials modelled as a continuous mass rather than as discrete particles. The French mathematician Augustin Louis Cauchy was the first to formulate such models in the 19th century, but research in the area continues today.

Modelling an object as a continuum assumes that the substance of the object completely fills the space it occupies. Modelling objects in this way ignores the fact that matter is made of atoms, and so is not continuous; however, on length scales much greater than that of inter-atomic distances, such models are highly accurate. Fundamental physical laws such as the conservation of mass, the conservation of momentum, and the conservation of energy may be applied to such models to derive differential equations describing the behaviour of such objects, and some information about the particular material studied is added through a constitutive relation.

Continuum mechanics deals with physical properties of solids and fluids which are independent of any particular coordinate system in which they are observed. These physical properties are then represented by tensors, which are mathematical objects that have the required property of being independent of coordinate system. These tensors can be expressed in coordinate systems for computational convenience.

The Concept of a Continuum

Materials, such as solids, liquids and gases, are composed of molecules separated by empty space. On a macroscopic scale, materials have cracks and discontinuities. However, certain physical phenomena can be modelled assuming the materials exist as a continuum, meaning the matter in the body is continuously distributed and fills the entire region of space it occupies. A continuum is a body that can be continually sub-divided into infinitesimal elements with properties being those of the bulk material.

The validity of the continuum assumption may be verified by a theoretical analysis, in which either some clear periodicity is identified or statistical homogeneity and ergodicity of the microstructure exists. More specifically, the continuum hypothesis/assumption hinges on the concepts of a *representative volume element* (RVE) (sometimes called "representative elementary volume") and *separation of scales* based on the Hill–Mandel condition. This condition provides a link between an experimentalist's and a theoretician's viewpoint on constitutive equations (linear and non-linear elastic/inelastic or coupled fields) as well as a way of spatial and statistical averaging of the microstructure.

When the separation of scales does not hold, or when one wants to establish a continuum of a finer resolution than that of the RVE size, one employs a *statistical volume element* (SVE), which, in turn, leads to random continuum fields. The latter then provide a micromechanics basis for stochastic finite elements (SFE). The levels of SVE and RVE link continuum mechanics to statistical mechanics. The RVE may be assessed only in a limited way via experimental testing: when the constitutive response becomes spatially homogeneous.

Specifically for fluids, the Knudsen number is used to assess to what extent the approximation of continuity can be made.

Formulation of Models

Continuum mechanics models begin by assigning a region in three dimensional Euclidean space to the material body being modelled. The points within this region are called particles or material points. Different *configurations* or states of the body correspond to different regions in Euclidean space. The region corresponding to the body's configuration at time t is $k_t(B)$ labelled.

A particular particle within the body in a particular configuration is characterised by a position vector

$$x = \sum_{i=1}^{3} x_i e_i ,$$

where e_1 are the coordinate vectors in some frame of reference chosen for the problem. This vector can be expressed as a function of the particle position X in some *reference configuration*, for example the configuration at the initial time, so that

$$x = k_t(X).$$

This function needs to have various properties so that the model makes physical sense. $k_t(.)$ needs to be:

- continuous in time, so that the body changes in a way which is realistic,
- globally invertible at all times, so that the body cannot intersect itself,
- orientation-preserving, as transformations which produce mirror reflections are not possible in nature.

For the mathematical formulation of the model, $k_t(.)$ is also assumed to be twice continuously differentiable, so that differential equations describing the motion may be formulated.

Forces in a Continuum

Continuum mechanics deals with deformable bodies, as opposed to rigid bodies. A solid is a deformable body that possesses shear strength, *sc.* a solid can support shear forces (forces parallel to the material surface on which they act). Fluids, on the other hand, do not sustain shear forces. For the study of the mechanical behaviour of solids and fluids these are assumed to be continuous bodies, which means that the matter fills the entire region of space it occupies, despite the fact that matter is made of atoms, has voids, and is discrete. Therefore, when continuum mechanics refers to a point or particle in a continuous body it does not describe a point in the interatomic space or an atomic particle, rather an idealised part of the body occupying that point.

Following the classical dynamics of Newton and Euler, the motion of a material body is produced by the action of externally applied forces which are assumed to be of two kinds: surface forces f_C and body forces f_B. Thus, the total force F applied to a body or to a portion of the body can be expressed as:

$$F = F_B + F_C$$

Surface forces or *contact forces*, expressed as force per unit area, can act either on the bounding surface of the body, as a result of mechanical contact with other bodies, or on imaginary internal surfaces that bound portions of the body, as a result of the mechanical interaction between the parts of the body to either side of the surface (Euler-Cauchy's stress principle). When a body is acted upon by external contact forces, internal contact forces are then transmitted from point to point inside the body to balance their action, according to Newton's second law of motion of conservation of linear momentum and angular momentum (for continuous bodies these laws are called the Euler's equations of motion). The internal contact forces are related to the body's deformation through constitutive equations. The internal contact forces may be mathematically described by how they relate to the motion of the body, independent of the body's material makeup.

Governing Equations

Continuum mechanics deals with the behaviour of materials that can be approximated as continuous for certain length and time scales. The equations that govern the mechanics of such materials include the balance laws for mass, momentum, and energy. Kinematic relations and constitutive equations are needed to complete the system of governing equations. Physical restrictions on the form of the constitutive relations can be applied by requiring that the second law of thermodynamics be satisfied under all conditions. In the continuum mechanics of solids, the second law of thermodynamics is satisfied if the Clausius–Duhem form of the entropy inequality is satisfied.

The balance laws express the idea that the rate of change of a quantity (mass, momentum, energy) in a volume must arise from three causes:

1. the physical quantity itself flows through the surface that bounds the volume,
2. there is a source of the physical quantity on the surface of the volume, or/and,
3. there is a source of the physical quantity inside the volume.

Let Ω be the body (an open subset of Euclidean space) and let be its surface (the boundary of Ω).

Let the motion of material points in the body be described by the map

$$x = \chi(X) = x(X)$$

where X is the position of a point in the initial configuration and is the location of the same point in the deformed configuration.

The deformation gradient is given by

$$F = \frac{\partial x}{\partial X} = x \cdot \nabla \ .$$

The Clausius–Duhem Inequality

The Clausius–Duhem inequality can be used to express the second law of thermodynamics for elastic-plastic materials. This inequality is a statement concerning the irreversibility of natural processes, especially when energy dissipation is involved.

Just like in the balance laws in the previous section, we assume that there is a flux of a quantity, a source of the quantity, and an internal density of the quantity per unit mass. The quantity of interest in this case is the entropy. Thus, we assume that there is an entropy flux, an entropy source, and an internal entropy density per unit mass (η) in the region of interest.

Let Ω be such a region and let $\partial\Omega$ be its boundary. Then the second law of thermodynamics states that the rate of increase of η in this region is greater than or equal to the sum of that supplied to Ω (as a flux or from internal sources) and the change of the internal entropy density due to material flowing in and out of the region.

Let $\partial\Omega$ move with a velocity u_n and let particles inside Ω have velocities V. Let n be the unit outward normal to the surface $\partial\Omega$. Let ρ be the density of matter in the region, q be the entropy flux at the surface, and r be the entropy source per unit mass. Then the entropy inequality may be written as

$$\frac{d}{dt}\left(\int_\Omega \rho\eta \, dV\right) \geq \int_{\partial\Omega} \rho\eta \, (u_n - v \cdot n) dA + \int_{\partial\Omega} \bar{q} \, dA + \int_\Omega \rho r \, dV.$$

The scalar entropy flux can be related to the vector flux at the surface by the relation $\bar{q} = -\psi(x) \cdot n$. Under the assumption of incrementally isothermal conditions, we have

$$\psi(x) = \frac{q(x)}{T} \ ; \ r = \frac{s}{T}$$

where q is the heat flux vector, s is a energy source per unit mass, and T is the absolute temperature of a material point x at at time t.

We then have the Clausius–Duhem inequality in integral form:

$$\frac{d}{dt}\left(\int_{\Omega}\rho\eta\, dV\right) \geq \int_{\partial\Omega}\rho\eta\,(u_n - v\cdot n)\, dA - \int_{\partial\Omega}\frac{q\cdot n}{T} dA + \int_{\Omega}\frac{\rho s}{T} dV.$$

We can show that the entropy inequality may be written in differential form as

$$\rho\dot{\eta} \geq -\nabla\cdot\left(\frac{q}{T}\right) + \frac{\rho s}{T}.$$

In terms of the Cauchy stress and the internal energy, the Clausius–Duhem inequality may be written as

$$\rho\,(\dot{e} - T\,\dot{\eta}) - \sigma : \nabla v \leq -\frac{q\cdot\nabla T}{T}.$$

Hydraulics

Hydraulics is a topic in applied science and engineering dealing with the mechanical properties of liquids. Fluid mechanics provides the theoretical foundation for hydraulics, which focuses on the engineering uses of fluid properties. In fluid power, hydraulics is used for the generation, control, and transmission of power by the use of pressurised liquids. Hydraulic topics range through most science and engineering disciplines, and cover concepts such as pipe flow, dam design, fluidics and fluid control circuitry, pumps, turbines, hydropower, computational fluid dynamics, flow measurement, river channel behaviour and erosion.

Free surface hydraulics is the branch of hydraulics dealing with free surface flow, such as occurring in rivers, canals, lakes, estuaries and seas. Its sub-field open channel flow studies the flow in open channels. The word "hydraulics" originates from the Greek word *hydraulikos* which in turn originates from *hydor*, Greek for water and *aulos*, meaning pipe.

Ancient and Medieval Ara

Early uses of water power date back to Mesopotamia and ancient Egypt, where irrigation has been used since the 6th millennium BC and water clocks had been used since the early 2nd millennium BC. Other early examples of water power include the Qanat system in ancient Persia and the Turpan water system in ancient China.

Greek/Hellenistic World

Greeks continued and sophisticated the construction of water and hydraulic power systems. An example is the construction by Eupalinos, under a public contract, of a watering channel for Samos. An early example of the usage of hydraulic wheel, probably the earliest in Europe, is the Perachora wheel (3rd c. BC).

Notable is the construction of the first hydraulic automata by Ctesibius (flourished c. 270 BC) and Hero of Alexandria (c. 10–80 AD). Hero describes a number of working machines using hydraulic power, such as the force pump, which is known from many Roman sites as having been used for raising water and in fire engines.

China

In ancient China there was Sunshu Ao (6th century BC), Ximen Bao (5th century BC), Du Shi (circa 31 AD), Zhang Heng (78 - 139 AD), and Ma Jun (200 - 265 AD), while medieval China had Su Song (1020 - 1101 AD) and Shen Kuo (1031–1095). Du Shi employed a waterwheel to power the bellows of a blast furnace producing cast iron. Zhang Heng was the first to employ hydraulics to provide motive power in rotating an armillary sphere for astronomical observation.

Sri Lanka

In ancient Sri Lanka, hydraulics were widely used in the ancient kingdoms of Anuradhapura and Polonnaruwa. The discovery of the principle of the valve tower, or valve pit, for regulating the escape of water is credited to ingenuity more than 2,000 years ago. By the first century A.D, several large-scale irrigation works had been completed. Macro- and micro-hydraulics to provide for domestic horticultural and agricultural needs, surface drainage and erosion control, ornamental and recreational water courses and retaining structures and also cooling systems were in place in Sigiriya, Sri Lanka. The coral on the massive rock at the site includes cisterns for collecting water.

Innovations in Ancient Rome

In Ancient Rome many different hydraulic applications were developed, including public water supplies, innumerable aqueducts, power using watermills and hydraulic mining. They were among the first to make use of the siphon to carry water across valleys, and used hushing on a large scale to prospect for and then extract metal ores. They used lead widely in plumbing systems for domestic and public supply, such as feeding thermae.

Hydraulic mining was used in the gold-fields of northern Spain, which was conquered by Augustus in 25 BC. The alluvial gold-mine of

Las Medulas was one of the largest of their mines. It was worked by at least 7 long aqueducts, and the water streams were used to erode the soft deposits, and then wash the tailings for the valuable gold content.

Modern Era (c. 1600–1870)

Benedetto Castelli: In 1619 Benedetto Castelli (1576 - 1578–1643), a student of Galileo Galilei, published the book *Della Misura dell'Acque Correnti* or "On the Measurement of Running Waters", one of the foundations of modern hydrodynamics. He served as a chief consultant to the Pope on hydraulic projects, i.e., management of rivers in the Papal States, beginning in 1626.

Blaise Pascal

Blaise Pascal (1623–1662-1672) studied fluid hydrodynamics and hydrostatics, centred on the principles of hydraulic fluids. His inventions include the hydraulic press, which multiplied a smaller force acting on a larger area into the application of a larger force totalled over a smaller area, transmitted through the same pressure (or same change of pressure) at both locations. Pascal's law or principle states that for an incompressible fluid at rest, the difference in pressure is proportional to the difference in height and this difference remains the same whether or not the overall pressure of the fluid is changed by applying an external force. This implies that by increasing the pressure at any point in a confined fluid, there is an equal increase at every other point in the container, i.e., any change in pressure applied at any point of the fluid is transmitted undiminished throughout the fluids.

Jean Louis Marie Poiseuille

A French physician, Poiseuille researched the flow of blood through the body and discovered an important law governing the rate of flow with the diameter of the tube in which flow occurred.

In the UK

Several cities developed city-wide hydraulic power networks in the 19th century, to operate machinery such as lifts, cranes, capstans and the like. Joseph Bramah was an early innovator and William Armstrong perfected the apparatus for power delivery on an industrial scale. In London, the London Hydraulic Power Company was a major supplier its pipes serving large parts of the West End of London, City and the Docks, but there were schemes restricted to single enterprises such as docks and railway goods yards.

7

Solid Mechanics

Solid mechanics is the branch of mechanics, physics, and mathematics that concerns the behaviour of solid matter under external actions (e.g., external forces, temperature changes, applied displacements, etc.). It is part of a broader study known as continuum mechanics. One of the most common practical applications of solid mechanics is the Euler-Bernoulli beam equation. Solid mechanics extensively uses tensors to describe stresses, strains, and the relationship between them.

Relationship to Continuum Mechanics

As shown in the following table, solid mechanics inhabits a central place within continuum mechanics. The field of rheology presents an overlap between solid and fluid mechanics.

Response Models

A material has a rest shape and its shape departs away from the rest shape due to stress. The amount of departure from rest shape is called deformation, the proportion of deformation to original size is called strain. If the applied stress is sufficiently low (or the imposed strain is small enough), almost all solid materials behave in such a way that the strain is directly proportional to the stress; the coefficient of the proportion is called the modulus of elasticity or Young's modulus. This region of deformation is known as the linearly elastic region.

It is most common for analysts in solid mechanics to use linear material models, due to ease of computation. However, real materials often exhibit non-linear behaviour. As new materials are used and old ones are pushed to their limits, non-linear material models are becoming more common.

There are three models that describe how a solid responds to an applied stress:

1. Elastically – When an applied stress is removed, the material returns to its undeformed state. Linearly elastic materials, those that deform proportionally to the applied load, can be described by the linear elasticity equations such as Hooke's law.
2. Viscoelastically – These are materials that behave elastically, but also have damping: when the stress is applied and removed, work has to be done against the damping effects and is converted in heat within the material resulting in a hysteresis loop in the stress–strain curve. This implies that the material response has time-dependence.
3. Plastically – Materials that behave elastically generally do so when the applied stress is less than a yield value. When the stress is greater than the yield stress, the material behaves plastically and does not return to its previous state. That is, deformation that occurs after yield is permanent.

Elasticity (Physics)

In physics, elasticity (or stretchiness) is the physical property of a material that returns to its original shape after the stress (e.g. external forces) that made it deform or distort is removed. The relative amount of deformation is called the strain.

The elastic regime is characterised by a linear relationship between stress and strain, denoted linear elasticity. The classic example is a metal spring. This idea was first stated by Robert Hooke in 1675 as a Latin anagram "ceiiinosssttuu" whose solution he published in 1678 as "*Ut tensio, sic vis*" which means "*As the extension, so the force.*"

This linear relationship is called Hooke's law. The classic model of linear elasticity is the perfect spring. Although the general proportionality constant between stress and strain in three dimensions is a 4th order tensor, when considering simple situations of higher symmetry such as a rod in one dimensional loading, the relationship may often be reduced to applications of Hooke's law.

Because most materials are elastic only under relatively small deformations, several assumptions are used to linearise the theory. Most importantly, higher order terms are generally discarded based on the small deformation assumption. In certain special cases, such as when considering a rubbery material, these assumptions may not be permissible. However, in general, elasticity refers to the linearised theory of the continuum stresses and strains.

Transitions to Inelasticity

A certain stress known as the elastic limit or the yield strength of an elastic material, the relationship between stress and strain becomes non-linear. Beyond this limit, the solid may deform irreversibly, exhibiting plasticity. A stress-strain curve is one tool for visualising this transition.

Not only solids exhibit elasticity. Some non-Newtonian fluids, such as viscoelastic fluids, will also exhibit elasticity in certain conditions. In response to a small, rapidly applied and removed strain, these fluids may deform and then return to their original shape. Under larger strains, or strains applied for longer periods of time, these fluids may start to flow like a liquid, with some viscosity.

Fracture Mechanics

Fracture mechanics is the field of mechanics concerned with the study of the propagation of cracks in materials. It uses methods of analytical solid mechanics to calculate the driving force on a crack and those of experimental solid mechanics to characterise the material's resistance to fracture.

In modern materials science, fracture mechanics is an important tool in improving the mechanical performance of materials and components. It applies the physics of stress and strain, in particular the theories of elasticity and plasticity, to the microscopic crystallographic defects found in real materials in order to predict the macroscopic mechanical failure of bodies. Fractography is widely used with fracture mechanics to understand the causes of failures and also verify the theoretical failure predictions with real life failures.

Linear Elastic Fracture Mechanics

Griffith's Criterion

Fracture mechanics was developed during World War I by English aeronautical engineer, A. A. Griffith, to explain the failure of brittle materials. Griffith's work was motivated by two contradictory facts:

- The stress needed to fracture bulk glass is around 100 MPa (15,000 psi).
- The theoretical stress needed for breaking atomic bonds is approximately 10,000 MPa (1,500,000 psi).

A theory was needed to reconcile these conflicting observations. Also, experiments on glass fibres that Griffith himself conducted suggested that the fracture stress increases as the fibre diameter decreases. Hence the uniaxial tensile strength, which had been used extensively to predict material failure before Griffith, could not be a specimen-independent material property. Griffith suggested that the low fracture strength observed in experiments, as well as the size-dependence of strength, was due to the presence of microscopic flaws in the bulk material.

To verify the flaw hypothesis, Griffith introduced an artificial flaw in his experimental specimens. The artificial flaw was in the form of a surface crack which was much larger than other flaws in a specimen. The experiments showed that the product of the square root of the flaw length (a) and the stress at fracture (σ_f) was nearly constant, which is expressed by the equation:

$$\sigma_f \sqrt{a} \approx C$$

An explanation of this relation in terms of linear elasticity theory is problematic. Linear elasticity theory predicts that stress (and hence the strain) at the tip of a sharp flaw in a linear elastic material is infinite. To avoid that problem, Griffith developed a thermodynamic approach to explain the relation that he observed.

The growth of a crack requires the creation of two new surfaces and hence an increase in the surface energy. Griffith found an expression for the constant C in terms of the surface energy of the crack by solving the elasticity problem of a finite crack in an elastic plate. Briefly, the approach was:

- Compute the potential energy stored in a perfect specimen under an uniaxial tensile load.
- Fix the boundary so that the applied load does no work and then introduce a crack into the specimen. The crack relaxes the stress and hence reduces the elastic energy near the crack faces. On the other hand, the crack increases the total surface energy of the specimen.
- Compute the change in the free energy (surface energy – elastic energy) as a function of the crack length. Failure occurs when the free energy attains a peak value at a critical crack length, beyond which the free energy decreases by increasing the crack length, i.e. by causing fracture. Using this procedure, Griffith found that

$$C = \sqrt{\frac{2E\gamma}{\pi}}$$

where E is the Young's modulus of the material and γ is the surface energy density of the material. Assuming E = 62 GPa and γ = 1 J/m^2 gives excellent agreement of Griffith's predicted fracture stress with experimental results for glass.

Irwin's Modification

Griffith's work was largely ignored by the engineering community until the early 1950s. The reasons for this appear to be (a) in the actual structural materials the level of energy needed to cause fracture is orders of magnitude higher than the corresponding surface energy, and (b) in structural materials there are always some inelastic deformations around the crack front that would make the assumption of linear elastic medium with infinite stresses at the crack tip highly unrealistic. F. Erdogan (2000)

Griffith's theory provides excellent agreement with experimental data for brittle materials such as glass. For ductile materials such as steel, though the relation $\sigma_y \sqrt{a} = C$ still holds, the surface energy (γ) predicted by Griffith's theory is usually unrealistically high. A group working under G. R. Irwin at the U.S. Naval Research Laboratory (NRL) during World War II realised that plasticity must play a significant role in the fracture of ductile materials.

In ductile materials (and even in materials that appear to be brittle), a plastic zone develops at the tip of the crack. As the applied load increases, the plastic zone increases in size until the crack grows and the material behind the crack tip unloads. The plastic loading and unloading cycle near the crack tip leads to the dissipation of energy as heat. Hence, a dissipative term has to be added to the energy balance relation devised by Griffith for brittle materials. In physical terms, additional energy is needed for crack growth in ductile materials when compared to brittle materials.

Irwin's strategy was to partition the energy into two parts:

- the stored elastic strain energy which is released as a crack grows. This is the thermodynamic driving force for fracture.
- the dissipated energy which includes plastic dissipation and the surface energy (and any other dissipative forces that may be at work). The dissipated energy provides the thermodynamic resistance to fracture. Then the total energy dissipated is

$$G = 2\gamma + G_p$$

where γ is the surface energy and G_p is the plastic dissipation (and dissipation from other sources) per unit area of crack growth.

The modified version of Griffith's energy criterion can then be written as

$$\sigma_f \sqrt{a} = \sqrt{\frac{EG}{\pi}}.$$

Stress Intensity Factor

Another significant achievement of Irwin and his colleagues was to find a method of calculating the amount of energy available for fracture in terms of the asymptotic stress and displacement fields around a crack front in a linear elastic solid. This asymptotic expression for the stress field around a crack tip is

$$\sigma_{ij} \approx \left(\frac{K}{\sqrt{2\pi r}}\right) f_{ij}(\theta)$$

where σ_{ij} are the Cauchy stresses, r is the distance from the crack tip, θ is the angle with respect to the plane of the crack, and f_{ij} are functions that are independent of the crack geometry and loading conditions. Irwin called the quantity K the *stress intensity factor*. Since the quantity f_{ij} is dimensionless, the stress intensity factor can be expressed in units of $\mathrm{MPa} - \sqrt{\mathrm{m}}$.

When a rigid line inclusion is considered, a similar asymptotic expression for the stress fields is obtained.

Strain Energy Release

Irwin was the first to observe that if the size of the plastic zone around a crack is small compared to the size of the crack, the energy required to grow the crack will not be critically dependent on the state of stress at the crack tip. In other words, a purely elastic solution may be used to calculate the amount of energy available for fracture.

The energy release rate for crack growth or *strain energy release rate* may then be calculated as the change in elastic strain energy per unit area of crack growth, i.e.,

$$G := -\left[\frac{\partial U}{\partial a}\right]_P = -\left[\frac{\partial U}{\partial a}\right]_u$$

where U is the elastic energy of the system and a is the crack length. Either the load P or the displacement u can be kept fixed while evaluating the above expressions.

Irwin showed that for a mode I crack (opening mode) the strain energy release rate and the stress intensity factor are related by:

$$G = G_I = \begin{cases} \frac{K_I^2}{E} & \text{plane stress} \\ \frac{(1-v^2)K_I^2}{E} & \text{plane strain} \end{cases}$$

where E is the Young's modulus, ν is Poisson's ratio, and K_I is the stress intensity factor in mode I. Irwin also showed that the strain energy release rate of a planar crack in a linear elastic body can be expressed in terms of the mode I, mode II (sliding mode), and mode III (tearing mode) stress intensity factors for the most general loading conditions.

Next, Irwin adopted the additional assumption that the size and shape of the energy dissipation zone remains approximately constant during brittle fracture. This assumption suggests that the energy needed to create a unit fracture surface is a constant that depends only on the material. This new material property was given the name *fracture toughness* and designated G_{Ic}. Today, it is the critical stress intensity factor K_{Ic} which is accepted as the defining property in linear elastic fracture mechanics.

Limitations

But a problem arose for the NRL researchers because naval materials, e.g., ship-plate steel, are not perfectly elastic but undergo significant plastic deformation at the tip of a crack. One basic assumption in Irwin's linear elastic fracture mechanics is that the size of the plastic zone is small compared to the crack length. However, this assumption is quite restrictive for certain types of failure in structural steels though such steels can be prone to brittle fracture, which has led to a number of catastrophic failures.

Linear-elastic fracture mechanics is of limited practical use for structural steels for another more practical reason. Fracture toughness testing is very expensive and engineers believe that sufficient information for selection of steels can be obtained from the simpler and cheaper Charpy impact test.

Elastic–Plastic Fracture Mechanics

Most engineering materials show some non-linear elastic and inelastic behaviour under operating conditions that involve large loads.

In such materials the assumptions of linear elastic fracture mechanics may not hold, that is,

- the plastic zone at a crack tip may have a size of the same order of magnitude as the crack size
- the size and shape of the plastic zone may change as the applied load is increased and also as the crack length increases.

Therefore a more general theory of crack growth is needed for elastic-plastic materials that can account for:

- the local conditions for initial crack growth which include the nucleation, growth, and coalescence of voids or decohesion at a crack tip.
- a global energy balance criterion for further crack growth and unstable fracture.

R-Curve

An early attempt in the direction of elastic-plastic fracture mechanics was Irwin's crack extension resistance curve or R-curve. This curve acknowledges the fact that the resistance to fracture increases with growing crack size in elastic-plastic materials. The R-curve is a plot of the total energy dissipation rate as a function of the crack size and can be used to examine the processes of slow stable crack growth and unstable fracture. However, the R-curve was not widely used in applications until the early 1970s. The main reasons appear to be that the R-curve depends on the geometry of the specimen and the crack driving force may be difficult to calculate.

J-Integral

In the mid-1960s James R. Rice (then at Brown University) and G. P. Cherepanov independently developed a new toughness measure to describe the case where there is sufficient crack-tip deformation that the part no longer obeys the linear-elastic approximation. Rice's analysis, which assumes non-linear elastic (or monotonic deformation theory plastic) deformation ahead of the crack tip, is designated the J integral. This analysis is limited to situations where plastic deformation at the crack tip does not extend to the furthest edge of the loaded part. It also demands that the assumed non-linear elastic behaviour of the material is a reasonable approximation in shape and magnitude to the real material's load response.

Cohesive Zone Models

When a significant region around a crack tip has undergone plastic deformation, other approaches can be used to determine the possibility of further crack extension and the direction of crack growth and branching. A simple technique that is easily incorporated into numerical calculations is the cohesive zone model method which is based on concepts proposed independently by Barenblatt and Dugdale in the early 1960s. The relationship between the Dugale-Barenblatt models and Griffith's theory was first discussed by Willis in 1967. The equivalence of the two approaches in the context of brittle fracture was shown by Rice in 1968. Interest in cohesive zone modelling of fracture has been reignited since 2000 following the pioneering work on dynamic fracture by Xu and Needleman , and Camacho and Ortiz.

Fully Plastic Failure

If the material is so tough that the yielded region ahead of the crack extends to the far edge of the specimen before fracture, the crack is no longer an effective stress concentrator. Instead, the presence of the crack merely serves to reduce the load-bearing area. In this regime the failure stress is conventionally assumed to be the average of the yield and ultimate strengths of the material.

Engineering Applications

The following information is needed for a fracture mechanics prediction of failure:

- Applied load
- Residual stress
- Size and shape of the part
- Size, shape, location, and orientation of the crack.

Usually not all of this information is available and conservative assumptions have to be made.

Occasionally post-mortem fracture-mechanics analyses are carried out. In the absence of an extreme overload, the causes are either insufficient toughness (K_{Ic}) or an excessively large crack that was not detected during routine inspection.

Short Summary

Arising from the manufacturing process, interior and surface flaws are found in all metal structures. Not all such flaws are unstable under service conditions. Fracture mechanics is the analysis of flaws to discover those that are safe (that is, do not grow) and those that are liable to propagate as cracks and so cause failure of the flawed structure. Ensuring safe operation of structure despite these inherent flaws is achieved through damage tolerance analysis. Fracture mechanics as a subject for critical study has barely been around for a century and thus is relatively new. There is a high demand for engineers with fracture mechanics expertise—particularly in this day and age where engineering failure is considered 'shocking' amongst the general public.

Appendix: Mathematical Relations

Griffith's Criterion

For the simple case of a thin rectangular plate with a crack perpendicular to the load Griffith's theory becomes:

$$G = \frac{\pi\sigma^2 a}{E} \tag{1}$$

where G is the strain energy release rate, σ is the applied stress, a is half the crack length, and E is the Young's modulus. The strain energy release rate can otherwise be understood as: *the rate at which energy is absorbed by growth of the crack.*

However, we also have that:

$$G_c = \frac{\pi\sigma_f^2 a}{E} \tag{2}$$

If $G \geq G_c$, this is the criterion for which the crack will begin to propagate.

Irwin's Modifications

Eventually a modification of Griffith's solids theory emerged from this work; a term called stress intensity replaced strain energy release rate and a term called fracture toughness replaced surface weakness energy. Both of these terms are simply related to the energy terms that Griffith used:

$$K_I = \sigma\sqrt{\pi a} \tag{3}$$

and

$$K_c = \sigma\sqrt{EG_c} \quad \text{(for plane stress)} \tag{4}$$

$$K_c = \sqrt{\frac{EG_c}{1-\nu^2}} \quad \text{(for plane strain)} \tag{5}$$

where K_I is the stress intensity, K_c the fracture toughness, and ν is Poisson's ratio. It is important to recognise the fact that fracture parameter K_c has different values when measured under plane stress and plane strain

Fracture occurs when $K_I \geq K_C$. For the special case of plane strain deformation, K_c becomes K_{Ic} and is considered a material property. The subscript I arises because of the different ways of loading a material to enable a crack to propagate. It refers to so-called "mode I" loading as opposed to mode II or III:

There are three ways of applying a force to enable a crack to propagate:

- Mode I crack – Opening mode (a tensile stress normal to the plane of the crack)
- Mode II crack – Sliding mode (a shear stress acting parallel to the plane of the crack and perpendicular to the crack front)
- Mode III crack – Tearing mode (a shear stress acting parallel to the plane of the crack and parallel to the crack front)

We must note that the expression for K_I in equation 3 will be different for geometries other than the center-cracked infinite plate, as discussed. Consequently, it is necessary to introduce a dimensionless correction factor, *Y*, in order to characterise the geometry. We thus have:

$$K_I = Y\sigma\sqrt{\pi a} \tag{6}$$

where *Y* is a function of the crack length and width of sheet given by:

$$Y\left(\frac{a}{W}\right) = \sqrt{\sec\left(\frac{\pi a}{W}\right)} \tag{7}$$

for a sheet of finite width *W* containing a through-thickness crack of length $2a$, or

$$Y\left(\frac{a}{W}\right) = 1.12 - \frac{0.41}{\sqrt{\pi}}\frac{a}{W} + \frac{18.7}{\sqrt{\pi}}\left(\frac{a}{W}\right)^2 - \ldots \tag{8}$$

for a sheet of finite width *W* containing a through-thickness edge crack of length *a*

Acoustics

Acoustics is the interdisciplinary science that deals with the study of all mechanical waves in gases, liquids, and solids including vibration, sound, ultrasound and infrasound. A scientist who works in the field of acoustics is an acoustician while someone working in the field of acoustics technology may be called an acoustical engineer. The application of acoustics can be seen in almost all aspects of modern society with the most obvious being the audio and noise control industries.

Hearing is one of the most crucial means of survival in the animal world, and speech is one of the most distinctive characteristics of human development and culture. So it is no surprise that the science of acoustics spreads across so many facets of our society—music, medicine, architecture, industrial production, warfare and more. Art, craft, science and technology have provoked one another to advance the whole, as in many other fields of knowledge. Lindsay's 'Wheel of Acoustics' is a well accepted overview of the various fields in acoustics.

The word "acoustic" is derived from the Greek word *akoustikos*, meaning "of or for hearing, ready to hear" and that from *akoustos*, "heard, audible", which in turn derives from the verb *akouo*, "I hear".

The Latin synonym is "sonic", after which the term sonics used to be a synonym for acoustics and later a branch of acoustics. After acousticians had extended their studies to frequencies above and below the audible range, it became conventional to identify these frequency ranges as "ultrasonic" and "infrasonic" respectively, while letting the word "acoustic" refer to the entire frequency range without limit.

History of Acoustics

Early Research in Acoustics

In the 6th century BC, the Greek philosopher Pythagoras wanted to know why some musical intervals seemed more beautiful than others, and he found answers in terms of numerical ratios representing the harmonic overtone series on a string. He is reputed to have observed that when the lengths of vibrating strings are expressible as ratios of integers (e.g. 2 to 3, 3 to 4), the tones produced will be harmonious. If, for example, a string sounds the note C when plucked, a string twice as long will sound the same note an octave lower. The tones in between are then given by 16:9 for D, 8:5 for E, 3:2 for F, 4:3 for G, 6:5 for A, and 16:15 for B, in ascending order. Aristotle (384-322 BC) understood that sound consisted of contractions and expansions of the air "falling upon and striking the air which is next to it...", a very good expression of the nature of wave motion. In about 20 BC, the Roman architect and engineer Vitruvius wrote a treatise on the acoustic properties of theatres including discussion of interference, echoes, and reverberation—the beginnings of architectural acoustics.

The physical understanding of acoustical processes advanced rapidly during and after the Scientific Revolution. Mainly Galileo Galilei (1564–1642) but also Marin Mersenne (1588–1648), independently, discovered the complete laws of vibrating strings (completing what Pythagoras and Pythagoreans had started 2000 years earlier). Galileo wrote "Waves are produced by the vibrations of a sonorous body, which spread through the air, bringing to the tympanum of the ear a stimulus which the mind interprets as sound", a remarkable statement that points to the beginnings of physiological and psychological acoustics. Experimental measurements of the speed of sound in air were carried out successfully between 1630 and 1680 by a number of investigators, prominently Mersenne. Meanwhile Newton (1642–1727) derived the relationship for wave velocity in solids, a cornerstone of physical acoustics (Principia, 1687).

The Age of Enlightenment and Onward

The eighteenth century saw major advances in acoustics as mathematicians applied the new techniques of calculus to elaborate theories of sound wave propagation. In the nineteenth century the major figures of mathematical acoustics were Helmholtz in Germany, who consolidated the field of physiological acoustics, and Lord Rayleigh in England, who combined the previous knowledge with his own copious contributions to the field in his monumental work "The Theory of Sound". Also in the 19th century, Wheatstone, Ohm, and Henry developed the analogy between electricity and acoustics.

The twentieth century saw a burgeoning of technological applications of the large body of scientific knowledge that was by then in place. The first such application was Sabine's groundbreaking work in architectural acoustics, and many others followed. Underwater acoustics was used for detecting submarines in the first World War. Sound recording and the telephone played important roles in a global transformation of society. Sound measurement and analysis reached new levels of accuracy and sophistication through the use of electronics and computing. The ultrasonic frequency range enabled wholly new kinds of application in medicine and industry. New kinds of transducers (generators and receivers of acoustic energy) were invented and put to use.

Fundamental Concepts of Acoustics

The study of acoustics revolves around the generation, propagation and reception of mechanical waves and vibrations.

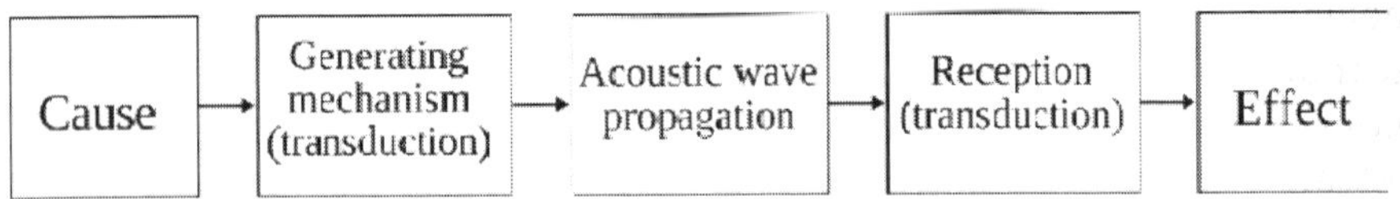

The steps shown in the above diagram can be found in any acoustical event or process. There are many kinds of cause, both natural and volitional. There are many kinds of transduction process that convert energy from some other form into acoustic energy, producing the acoustic wave. There is one fundamental equation that describes acoustic wave propagation, but the phenomena that emerge from it are varied and often complex. The wave carries energy throughout the propagating medium. Eventually this energy is transduced again into other forms, in ways that again may be natural and/or volitionally contrived. The final effect may be purely physical or it may reach far into the biological or volitional domains. The five basic steps are found equally well whether we are talking about an earthquake, a submarine using sonar to locate its foe, or a band playing in a rock concert.

The central stage in the acoustical process is wave propagation. This falls within the domain of physical acoustics. In fluids, sound propagates primarily as a pressure wave. In solids, mechanical waves can take many forms including longitudinal waves, transverse waves and surface waves.

Acoustics looks first at the pressure levels and frequencies in the sound wave. Transduction processes are also of special importance.

Wave Propagation: Pressure Levels

In fluids such as air and water, sound waves propagate as disturbances in the ambient pressure level. While this disturbance is usually small, it is still noticeable to the human ear. The smallest sound that a person can hear, known as the threshold of hearing, is nine orders of magnitude smaller than the ambient pressure. The loudness of these disturbances is called the sound pressure level (SPL), and is measured on a logarithmic scale in decibels.

Wave Propagation: Frequency

Physicists and acoustic engineers tend to discuss sound pressure levels in terms of frequencies, partly because this is how our ears interpret sound. What we experience as "higher pitched" or "lower pitched" sounds are pressure vibrations having a higher or lower number of cycles per second. In a common technique of acoustic measurement, acoustic signals are sampled in time, and then presented in more meaningful forms such as octave bands or time frequency plots. Both these popular methods are used to analyse sound and better understand the acoustic phenomenon.

The entire spectrum can be divided into three sections: audio, ultrasonic, and infrasonic. The audio range falls between 20 Hz and 20,000 Hz. This range is important because its frequencies can be detected by the human ear. This range has a number of applications, including speech communication and music. The ultrasonic range refers to the very high frequencies: 20,000 Hz and higher. This range has shorter wavelengths which allows better resolution in imaging technologies. Medical applications such as ultrasonography and elastography rely on the ultrasonic frequency range. On the other end of the spectrum, the lowest frequencies are known as the infrasonic range. These frequencies can be used to study geological phenomena such as earthquakes.

Analytic instruments such as the Spectrum analyser facilitate visualisation and measurement of acoustic signals and their properties. The Spectrogram produced by such an instrument is a graphical display of the time varying pressure level and frequency profiles which give a specific acoustic signal its defining character.

Transduction in Acoustics

A transducer is a device for converting one form of energy into another. In an acoustical context, this usually means converting sound energy into electrical energy (or vice versa). For nearly all acoustic applications, some type of acoustic transducer is necessary. Acoustic transducers include loudspeakers, microphones, hydrophones and sonar projectors. These devices convert an electric signal to or from a sound pressure wave. The most widely used transduction principles are electromagnetism, electrostatics and piezoelectricity.

The transducers in most common loudspeakers (e.g. woofers and tweeters), are electromagnetic devices that generate waves using a suspended diaphragm driven by an electromagnetic voice coil, sending off pressure waves. Electret microphones and condenser microphones employ electrostatics. As the sound wave strikes the microphone's diaphragm, it moves and induces a voltage change. The ultrasonic systems used in medical ultrasonography employ piezoelectric transducers. These are made from special ceramics in which elastic vibrations and electrical fields are interlinked through a property of the material itself.

Statics

Statics is the branch of mechanics concerned with the analysis of loads (force, torque/moment) on physical systems in static equilibrium, that is, in a state where the relative positions of subsystems do not vary over time, or where components and structures are at a constant velocity. When in static equilibrium, the system is either at rest, or its center of mass moves at constant velocity.

By Newton's first law, this situation implies that the net force and net torque (also known as moment of force) on every body in the system is zero. From this constraint, such quantities as stress or pressure can be derived. The net forces equalling zero is known as the *first condition for equilibrium,* and the net torque equalling zero is known as the *second condition for equilibrium.*

Vectors

Example of a beam in static equilibrium. The sum of force and moment is zero.

A scalar is a quantity like mass or temperature which only has a magnitude. A vector is a quantity that has both a magnitude and a direction. There are many notations to identify a vector, the most common ones are:

- A bold faced character V
- An underlined character V
- A character with an arrow over it $\vec{V}$.

Vectors can be added using the parallelogram law or the triangle law. Vectors contain components in orthogonal bases. Unit vectors i, j, and k are by convention along the x, y, and z directions.

Force

Force is the action of one body on another. A force tends to move a body in the direction of its action. The action of a force is characterised by its magnitude, by the direction of its action, and by its point of application. Thus force is a vector quantity, because its effect depends on the direction as well as on the magnitude of the action. (Meriam 2007 p. 23)

Forces are classified as either contact or body forces. A contact force is produced by direct physical contact; an example is the force exerted on a body by a supporting surface. On the other hand, a body force is generated by virtue of the position of a body within a force field such as a gravitational, electric, or magnetic field. An example of a body force is your weight. (Meriam 2007 p. 24)

Moment of a Force

In addition to the tendency to move a body in the direction of its application, a force can also tend to rotate a body about an axis. The axis may be any line which neither intersects nor is parallel to the line of action of the force. This rotational tendency is known as the *moment* M of the force. Moment is also referred to as *torque.*

Moment About a Point

The magnitude of the moment of a force at a point O, is equal to the perpendicular distance from O to the line of action of F, multiplied by the magnitude of the force. Simply the magnitude of the moment is defined as

$$M = Fd$$

where

F = The force applied

d = The perpendicular distance from the axis to the line of action of the force. Normally referred to as the moment arm.

The direction of the moment is given by the right hand rule, where counter clockwise (CCW) is out of the page, and clockwise (CW) is into the page. The moment direction may be accounted for by using a stated sign convention, such as a plus sign (+) for counterclockwise moments and a minus sign (-) for clockwise moments, or vice versa. Moments can be added together as vectors.

Varignon's Theorem

One of the most useful principles of mechanics is Varignon's theorem which states that the moment of a force about any point is equal to the sum of the moments of the components of the force about the same point.

Equilibrium Equations

The static equilibrium of a particle is an important concept in Statics. A particle is in equilibrium only if the resultant of all forces acting on the particle is equal to zero. In a rectangular coordinate system the equilibrium equations can be represented by three scalar equations, where the sum of forces in all three directions are equal to zero. An engineering application of this concept is determining the tensions of up to three cables under load, for example the forces exerted on each cable of a hoist lifting an object or of guy wires restraining a hot air balloon to the ground.

Moment of Inertia

In classical mechanics, moment of inertia, also called mass moment, rotational inertia, polar moment of inertia of mass, or the angular mass, (SI units kg·m^2) is a measure of an object's resistance to changes to its rotation. It is the inertia of a rotating body with respect to its rotation. The moment of inertia plays much the same role in rotational dynamics as mass does in linear dynamics, describing the relationship between angular momentum and angular velocity, torque and angular acceleration, and several other quantities. The symbol I and sometimes J are usually used to refer to the moment of inertia or polar moment of inertia.

While a simple scalar treatment of the moment of inertia suffices for many situations, a more advanced tensor treatment allows the analysis of such complicated systems as spinning tops and gyroscopic motion.

The concept was introduced by Leonhard Euler in his book *Theoria motus corporum solidorum seu rigidorum* in 1765. In this book, he discussed the moment of inertia and many related concepts, such as the principal axis of inertia.

Solids

Statics is used in the analysis of structures, for instance in architectural and structural engineering. Strength of materials is a related field of mechanics that relies heavily on the application of static equilibrium. A key concept is the center of gravity of a body at rest: it represents an imaginary point at which all the mass of a body resides. The position of the point relative to the foundations on which a body lies determines its stability towards small movements. If the center of gravity exists outside the foundations, then the body is unstable because there is a torque acting: any small disturbance will cause the body to fall or topple. If the center of gravity exists within the foundations, the body is stable since no net torque acts on the body. If the center of gravity coincides with the foundations, then the body is said to be metastable.

Fluids

Hydrostatics, also known as fluid statics, is the study of fluids at rest. This analyses bodies of fluid in static equilibrium. The characteristic of any fluid at rest is that the force exerted on any particle of the fluid is the same at all points at the same depth (or altitude) within the fluid. If the net force is greater than zero the fluid will move in the direction of the resulting force. This concept was first formulated in a slightly extended form by the French mathematician and philosopher Blaise Pascal in 1647 and would be later known as Pascal's Law. This law has many important applications in hydraulics. Archimedes, Abu Rayhan al-Biruni, Al-Khazini and Galileo Galilei were also major figures in the development of hydrostatics.

Mechanical Equilibrium

A standard definition of static equilibrium is:

A system of particles is in static equilibrium when all the particles of the system are at rest and the total force on each particle is permanently zero.

This is a strict definition, and often the term "static equilibrium" is used in a more relaxed manner interchangeably with "mechanical equilibrium", as defined next.

A standard definition of mechanical equilibrium for a particle is:

The necessary and sufficient conditions for a particle to be in mechanical equilibrium is that the net force acting upon the particle is zero.

The necessary conditions for mechanical equilibrium for a system of particles are:

(i) The vector sum of all *external forces* is zero;

(ii) The sum of the moments of all *external forces* about any line is zero.

As applied to a rigid body, the necessary and sufficient conditions become:

A rigid body is in mechanical equilibrium when the sum of all forces on all particles of the system is zero, and also the sum of all torques on all particles of the system is zero.

A rigid body in mechanical equilibrium is undergoing neither linear nor rotational acceleration; however it could be translating or rotating at a constant velocity.

However, this definition is of little use in continuum mechanics, for which the idea of a particle is foreign. In addition, this definition gives no information as to one of the most important and interesting aspects of equilibrium states – their stability.

An alternative definition of equilibrium that applies to conservative systems and often proves more useful is:

A system is in mechanical equilibrium if its position in configuration space is a point at which the gradient with respect to the generalised coordinates of the potential energy is zero.

Because of the fundamental relationship between force and energy, this definition is equivalent to the first definition. However, the definition involving energy can be readily extended to yield information about the stability of the equilibrium state.

For example, from elementary calculus, we know that a necessary condition for a local minimum *or* a maximum of a differentiable function is a vanishing first derivative (that is, the first derivative is becoming zero). To determine whether a point is a minimum or maximum, one may be able to use the second derivative test. The consequences to the stability of the equilibrium state are as follows:

- Second derivative < 0 : The potential energy is at a local maximum, which means that the system is in an unstable equilibrium state. If the system is displaced an arbitrarily small distance from the equilibrium state, the forces of the system cause it to move even farther away.
- Second derivative > 0 : The potential energy is at a local minimum. This is a stable equilibrium. The response to a small perturbation is forces that tend to restore the equilibrium. If more than one stable equilibrium state is possible for a system, any equilibria whose potential energy is higher than the absolute minimum represent metastable states.
- Second derivative $= 0$ or does not exist: The second derivative test fails, and one must typically resort to using the first derivative test. Both of the previous results are still possible, as is a third: this could be a region in which the energy does not vary, in which case the equilibrium is called neutral or indifferent or marginally stable. To lowest order, if the system is displaced a small amount, it will stay in the new state.

In more than one dimension, it is possible to get different results in different directions, for example stability with respect to displacements in the x-direction but instability in the y-direction, a case known as a saddle point. Without further qualification, an equilibrium is stable only if it is stable in all directions.

The special case of mechanical equilibrium of a stationary object is static equilibrium. A paperweight on a desk would be in static equilibrium. The minimal number of static equilibria of homogeneous, convex bodies (when resting under gravity on a horizontal surface) is of special interest. In the planar case, the minimal number is 4, while in three dimensions one can build an object with just one stable and one unstable balance point, this is called Gomboc. A child sliding down a slide at constant speed would be in mechanical equilibrium, but not in static equilibrium.

An example of mechanical equilibrium is a person trying to press a spring. He or she can push it up to a point after which it reaches a state where the force trying to compress it and the resistive force from the spring are equal, so the person cannot further press it. At this state the system will be in mechanical equilibrium. When the pressing force is removed the spring attains its original state.

8

Structural Analysis

Structural analysis is the determination of the effects of loads on physical structures and their components. Structures subject to this type of analysis include all that must withstand loads, such as buildings, bridges, vehicles, machinery, furniture, attire, soil strata, prostheses and biological tissue. Structural analysis incorporates the fields of applied mechanics, materials science and applied mathematics to compute a structure's deformations, internal forces, stresses, support reactions, accelerations, and stability. The results of the analysis are used to verify a structure's fitness for use, often saving physical tests. Structural analysis is thus a key part of the engineering design of structures.

Structures and Loads

A structure refers to a body or system of connected parts used to support a load. Important examples related to Civil Engineering include buildings, bridges, and towers; and in other branches of engineering, ship and aircraft frames, tanks, pressure vessels, mechanical systems, and electrical supporting structures are important. In order to design a structure, one must serve a specified function for public use, the engineer must account for its safety, aesthetics, and serviceability, while taking into consideration economic and environmental constraints. Other branches of engineering work on a wide variety of nonbuilding structures.

Classification of Structures

It is important for a structural engineer to recognise the various types of elements composing a structure and to be able to classify structures as to their form and function. Some of the structural elements are tie rods, rod, bar, angle, channel, beams, and columns. Combination of structural elements and the materials from which they are composed is referred to as a structural system. Each system is constructed of one or more basic types of structures such as Trusses, Cables and Arches, Frames, and Surface Structures.

Loads

Once the dimensional requirement for a structure have been defined, it becomes necessary to determine the loads the structure must support. In order to design a structure, it is therefore necessary to first specify the loads that act on it. The design loading for a structure is often specified in building codes. There are two types of codes: general building codes and design codes, engineer must satisfy all the codes requirements for a reliable structure. There are two types of loads that structure engineering must encounter in the design. First type of load is called Dead loads that consist of the weights of the various structural members and the weights of any objects that are permanently attached to the structure. For example, columns, beams, girders, the floor slab, roofing, walls, windows, plumbing, electrical fixtures, and other miscellaneous attachments. Second type of load is Live Loads which vary in their magnitude and location. There are many different types of live loads like building loads, highway bridge Loads, railroad bridge Loads, impact loads, wind loads, snow loads, earthquake loads, and other natural loads.

Analytical Methods

To perform an accurate analysis a structural engineer must determine such information as structural loads, geometry, support conditions, and materials properties. The results of such an analysis typically include support reactions, stresses and displacements. This information is then compared to criteria that indicate the conditions of failure. Advanced structural analysis may examine dynamic response, stability and non-linear behaviour.

There are three approaches to the analysis: the mechanics of materials approach (also known as strength of materials), the elasticity theory approach (which is actually a special case of the more general field of continuum mechanics), and the finite element approach. The first two make use of analytical formulations which apply mostly to simple linear elastic models, lead to closed-form solutions, and can often be solved by hand. The finite element approach is actually a numerical method for solving differential equations generated by theories of mechanics such as elasticity theory and strength of materials. However, the finite-element method depends heavily on the processing power of computers and is more applicable to structures of arbitrary size and complexity.

Regardless of approach, the formulation is based on the same three fundamental relations: equilibrium, constitutive, and compatibility. The solutions are approximate when any of these relations are only approximately satisfied, or only an approximation of reality.

Limitations

Each method has noteworthy limitations. The method of mechanics of materials is limited to very simple structural elements under relatively simple loading conditions. The structural elements and loading conditions allowed, however, are sufficient to solve many useful engineering problems. The theory of elasticity allows the solution of structural elements of general geometry under general loading conditions, in principle. Analytical solution, however, is limited to relatively simple cases. The solution of elasticity problems also requires the solution of a system of partial differential equations, which is considerably more mathematically demanding than the solution of mechanics of materials problems, which require at most the solution of an ordinary differential equation. The finite element method is perhaps the most restrictive and most useful at the same time. This method itself relies upon other structural theories (such as the other two discussed here) for equations to solve. It does, however, make it generally possible to solve these equations, even with highly complex geometry and loading conditions, with the restriction that there is always some numerical error. Effective and reliable use of this method requires a solid understanding of its limitations.

Strength of Materials Methods (Classical Methods)

The simplest of the three methods here discussed, the mechanics of materials method is available for simple structural members subject to specific loadings such as axially loaded bars, prismatic beams in a state of pure bending, and circular shafts subject to torsion. The solutions can under certain conditions be superimposed using the superposition principle to analyse a member undergoing combined loading. Solutions for special cases exist for common structures such as thin-walled pressure vessels.

For the analysis of entire systems, this approach can be used in conjunction with statics, giving rise to the *method of sections* and *method of joints* for truss analysis, moment distribution method for small rigid frames, and *portal frame* and *cantilever method* for large rigid frames. Except for moment distribution, which came into use in the 1930s, these methods were developed in their current forms in the second half of the nineteenth century. They are still used for small structures and for preliminary design of large structures.

The solutions are based on linear isotropic infinitesimal elasticity and Euler-Bernoulli beam theory. In other words, they contain the assumptions (among others) that the materials in question are elastic, that stress is related linearly to strain, that the material (but not the structure) behaves identically regardless of direction of the applied load, that all deformations are small, and that beams are long relative to their depth. As with any simplifying assumption in engineering, the more the model strays from reality, the less useful (and more dangerous) the result.

Elasticity Methods

Elasticity methods are available generally for an elastic solid of any shape. Individual members such as beams, columns, shafts, plates and shells may be modelled. The solutions are derived from the equations of linear elasticity. The equations of elasticity are a system of 15 partial differential equations. Due to the nature of the mathematics involved, analytical solutions may only be produced for relatively simple geometries. For complex geometries, a numerical solution method such as the finite element method is necessary.

Many of the developments in the mechanics of materials and elasticity approaches have been expounded or initiated by Stephen Timoshenko.

Methods Using Numerical Approximation

It is common practice to use approximate solutions of differential equations as the basis for structural analysis. This is usually done using numerical approximation techniques. The most commonly used numerical approximation in structural analysis is the Finite Element Method.

The finite element method approximates a structure as an assembly of elements or components with various forms of connection between them. Thus, a continuous system such as a plate or shell is modelled as a discrete system with a finite number of elements interconnected at finite number of nodes. The behaviour of individual elements is characterised by the element's stiffness or flexibility relation, which altogether leads to the system's stiffness or flexibility relation. To establish the element's stiffness or flexibility relation, we can use the *mechanics of materials* approach for simple one-dimensional bar elements, and the *elasticity approach* for more complex two- and three-dimensional elements. The analytical and computational development are best effected throughout by means of matrix algebra.

Early applications of matrix methods were for articulated frameworks with truss, beam and column elements; later and more advanced matrix methods, referred to as "finite element analysis," model an entire structure with one-, two-, and three-dimensional elements and can be used for articulated systems together with continuous systems such as a pressure vessel, plates, shells, and three-dimensional solids. Commercial computer software for structural analysis typically uses matrix finite-element analysis, which can be further classified into two main approaches: the displacement or stiffness method and the force or flexibility method. The stiffness method is the most popular by far thanks to its ease of implementation as well as of formulation for advanced applications. The finite-element technology is now sophisticated enough to handle just about any system as long as sufficient computing power is available. Its applicability includes, but is not limited to, linear and non-linear analysis, solid and fluid interactions, materials that are isotropic, orthotropic, or anisotropic, and external effects that are static, dynamic, and environmental factors. This, however, does not imply that the computed solution will automatically be reliable because much depends on the model and the reliability of the data input.

Time-Line

- 1452-1519 Leonardo da Vinci made many contributions.
- 1638: Galileo Galilei published the book "Two New Sciences" in which he examined the failure of simple structures.
- 1660: Hooke's law by Robert Hooke.
- 1687: Isaac Newton published "Philosophiae Naturalis Principia Mathematica" which contains the Newton's laws of motion.
- 1750: Euler-Bernoulli beam equation.
- 1700-1782: Daniel Bernoulli introduced the principle of virtual work.
- 1707-1783: Leonhard Euler developed the theory of buckling of columns.
- 1826: Claude-Louis Navier published a treatise on the elastic behaviours of structures.

- 1873: Carlo Alberto Castigliano presented his dissertation "Intorno ai sistemi elastici", which contains his theorem for computing displacement as partial derivative of the strain energy. This theorem includes the method of *least work* as a special case.
- 1936: Hardy Cross' publication of the moment distribution method which was later recognised as a form of the relaxation method applicable to the problem of flow in pipe-network.
- 1941: Alexander Hrennikoff submitted his D.Sc thesis in MIT on the discretisation of plane elasticity problems using a lattice framework.
- 1942: R. Courant divided a domain into finite subregions.
- 1956: J. Turner, R. W. Clough, H. C. Martin, and L. J. Topp's paper on the "Stiffness and Deflection of Complex Structures". This paper introduces the name "finite-element method" and is widely recognised as the first comprehensive treatment of the method as it is known today.

Failure Analysis

Failure analysis is the process of collecting and analysing data to determine the cause of a failure. It is an important discipline in many branches of manufacturing industry, such as the electronics industry, where it is a vital tool used in the development of new products and for the improvement of existing products. It relies on collecting failed components for subsequent examination of the cause or causes of failure using a wide array of methods, especially microscopy and spectroscopy. The NDT or nondestructive testing methods are valuable because the failed products are unaffected by analysis, so inspection always starts using these methods.

Forensic Investigation

Forensic enquiry into the failed process or product is the starting point of failure analysis. Such enquiry is conducted using scientific analytical methods such as electrical and mechanical measurements, or by analysing failure data such as product reject reports or examples of previous failures of the same kind. The methods of forensic engineering are especially valuable in tracing product defects and flaws. They may include fatigue cracks, brittle cracks produced by stress corrosion cracking or environmental stress cracking for example. Witness statements can be valuable for reconstructing the likely sequence of events and hence the chain of cause and effect. Human factors can also be assessed when the cause of the failure is determined. There are several useful methods to prevent product failures occurring in the first place, including failure mode and effects analysis (FMEA) and fault tree analysis (FTA), methods which can be used during prototyping to analyse failures before a product is marketed.

Failure theories can only be constructed on such data, but when corrective action is needed quickly, the precautionary principle demands that measures be put in place. In aircraft accidents for example, all planes of the type involved can be grounded immediately pending the outcome of the enquiry.

Another aspect of failure analysis is associated with No Fault Found (NFF) which is a term used in the field of failure analysis to describe a situation where an originally reported mode of failure can't be duplicated by the evaluating technician and therefore the potential defect can't be fixed.

NFF can be attributed to oxidation, defective connections of electrical components, temporary shorts or opens in the circuits, software bugs, temporary environmental factors, but also to the operator error. Large number of devices that are reported as NFF during the first troubleshooting session often return to the failure analysis lab with the same NFF symptoms or a permanent mode of failure.

The term Failure analysis also applies to other fields such as business management and military strategy.

Failure Analysis Engineers

A failure analysis engineer often plays a lead role in the analysis of failures, whether a component or product fails in service or if failure occurs in manufacturing or during production processing. In any case, one must determine the cause of failure to prevent future occurrence, and/or to improve the performance of the device, component or structure.

Methods of Analysis

The failure analysis of many different products involves the use of the following tools and techniques:

Microscopes

- Optical microscope
  - Liquid crystal
- Scanning acoustic microscope (SAM)

- Scanning Acoustic Tomography (SCAT)
- Atomic Force Microscope (AFM)
- Stereomicroscope
- Photo emission microscope (PEM)
- X-ray microscope
- Infra-red microscope
- Scanning SQUID microscope.

Sample Preparation

- Jet-etcher
- Plasma etcher
- Back Side Thinning Tools
    - Mechanical Back Side Thinning
    - Laser Chemical Back Side Etching.

Spectroscopic Analysis

- Transmission line pulse spectroscopy (TLPS)
- Auger electron spectroscopy
- Deep Level Transient Spectroscopy (DLTS).

Device Modification

- Focused ion beam etching (FIB).

Surface Analysis

- Dye penetrant inspection.

Scanning Electron Microscopy

- Scanning electron microscope (SEM)
    - Electron beam induced current (EBIC) in SEM
    - Charge Induced Voltage Alteration (CIVA) in SEM
    - Voltage contrast in SEM
    - Electron backscatter diffraction (EBSD) in SEM
    - Energy Dispersive X-ray Spectroscopy (EDS) in SEM
- Transmission electron microscope (TEM).

Laser Signal Injection Microscopy (LSIM)

- Photo carrier stimulation
    - Static
        - Optical beam induced current (OBIC)

        - Light Induced Voltage Alteration (LIVA)
    - Dynamic
        - Laser Assisted Device Alteration (LADA)
- Thermal Laser Stimulation (TLS)
    - Static
        - Optical Beam Induced Resistance Change (OBIRCH)
        - Thermally Induced Voltage Alteration (TIVA)
        - External Induced Voltage Alteration (XIVA)
        - Seebeck Effect Imaging (SEI)
    - Dynamic
        - Soft Defect Localisation (SDL).

Semiconductor Probing

- Mechanical Probe Station
- Electron Beam Prober
- Laser Voltage Prober
- Time-Resolved Photon Emission Prober (TRPE)

Software Based Fault Location Techniques

- CAD Navigation
- Automatic Test Pattern Generation (ATPG)

This list is incomplete; you can help by expanding it.

Thermodynamics

Thermodynamics is a physical science that studies the effects on material bodies, and on radiation in regions of space, of transfer of heat and of work done on or by the bodies or radiation. It interrelates macroscopic variables, such as temperature, volume and pressure, which describe physical properties of material bodies and radiation, which in this science are called thermodynamic systems.

Historically, thermodynamics developed out of a desire to increase the efficiency of early steam engines, particularly through the work of French physicist Nicolas Léonard Sadi Carnot (1824) who believed that the efficiency of heat engines was the key that could help France win the Napoleonic Wars. Scottish physicist Lord Kelvin was the first to formulate a concise definition of thermodynamics in 1854:

Thermo-dynamics is the subject of the relation of heat to forces acting between contiguous parts of bodies, and the relation of heat to electrical agency.

Initially, the thermodynamics of heat engines concerned mainly the thermal properties of their 'working materials', such as steam. This concern was then linked to the study of energy transfers in chemical processes, for example to the investigation, published in 1840, of the heats of chemical reactions by Germain Hess, which was not originally explicitly concerned with the relation between energy exchanges by heat and work. Chemical thermodynamics studies the role of entropy in chemical reactions. Also, statistical thermodynamics, or statistical mechanics, gave explanations of macroscopic thermodynamics by statistical predictions of the collective motion of particles based on the mechanics of their microscopic behaviour.

Thermodynamics describes how systems change when they interact with one another or with their surroundings. This can be applied to a wide variety of topics in science and engineering, such as engines, phase transitions, chemical reactions, transport phenomena, and even black holes. The results of thermodynamics are essential for other fields of physics and for chemistry, chemical engineering, aerospace engineering, mechanical engineering, cell biology, biomedical engineering, materials science, and are useful for other fields such as economics.

Many of the empirical facts of thermodynamics are comprehended in its four laws. The first law specifies that energy can be exchanged between physical systems as heat and thermodynamic work. The second law concerns a quantity called entropy, that expresses limitations, arising from what is known as irreversibility, on the amount of thermodynamic work that can be delivered to an external system by a thermodynamic process. Many writers offer various axiomatic formulations of thermodynamics, as if it were a completed subject, but non-equilibrium processes continue to make difficulties for it.

Introduction

Thermodynamics is built on the study of energy transfers that can be strictly resolved into two distinct components, heat and work, specified by macroscopic variables. Though thermodynamics originated in the study of cyclic non-equilibrium processes such as the working of heat engines, study of the subject gradually revealed that the notion of heat is inextricably tied to the notion of thermodynamic equilibrium. Thermodynamics is well understood and validated for systems in thermodynamic equilibrium, but as the systems of interest become further and further from thermodynamic equilibrium, their thermodynamical study becomes more and more difficult. Systems in thermodynamic equilibrium have very well experimentally reproducible behaviour, and as interest moves further towards non-equilibrium systems, experimental reproducibility becomes more difficult. A gradual approach to the subject, starting with a focus on cyclic processes and thermodynamic equilibrium, and then gradually beginning to further consider non-equilibrium systems.

Basic for thermodynamics are the concepts of *system* and *surroundings.*

A thermodynamic system can be defined in terms of its states. In this way, a thermodynamic system is a macroscopic physical object, explicitly specified in terms of macroscopic physical and chemical variables which describe its macroscopic properties. The macroscopic state variables of thermodynamics have been recognised in the course of empirical work in physics and chemistry.

A thermodynamic system can also be defined in terms of the processes which it can undergo. Of particular interest are cyclic processes. This was the way of the founders of thermodynamics in the first three quarters of the nineteenth century.

The surroundings of a thermodynamic system are other thermodynamic systems that can interact with it. An example of a thermodynamic surrounding is a heat bath, which is considered to be held at a prescribed temperature, regardless of the interactions it might have with the system.

The macroscopic variables of a thermodynamic system can under some conditions be related to one another through equations of state. They express the constitutive peculiarities of the material of the system. Classical thermodynamics is characterised by its study of materials that have equations of state that express relations between mechanical variables and temperature that are reached much more rapidly than any changes in the surroundings. A classical material can usually be described by a function that makes pressure dependent on volume and temperature, the resulting pressure being established much more rapidly than any imposed change of volume or temperature.

Thermodynamic facts can often be explained by viewing macroscopic objects as assemblies of very many microscopic or atomic objects that obey Hamiltonian dynamics. The microscopic or atomic objects exist in species, the objects of each species being all alike. Because of this likeness, statistical methods can be used to account for the macroscopic properties of the thermodynamic system in terms of the properties of the microscopic species. Such explanation is called statistical thermodynamics; also often it is also referred to by the term 'statistical mechanics', though this term can have a wider meaning, referring to 'microscopic objects', such as economic quantities, that do not obey Hamiltonian dynamics.

History

The history of thermodynamics as a scientific discipline generally begins with Otto von Guericke who, in 1650, built and designed the world's first vacuum pump and demonstrated a vacuum using his Magdeburg hemispheres. Guericke was driven to make a vacuum in order to disprove Aristotle's long-held supposition that 'nature abhors a vacuum'. Shortly after Guericke, the English physicist and chemist Robert Boyle had learned of Guericke's designs and, in 1656, in coordination with English scientist Robert Hooke, built an air pump. Using this pump, Boyle and Hooke noticed a correlation between pressure, temperature, and volume. In time, Boyle's Law was formulated, which states that pressure and volume are inversely proportional. Then, in 1679, based on these concepts, an associate of Boyle's named Denis Papin built a steam digester, which was a closed vessel with a tightly fitting lid that confined steam until a high pressure was generated.

Later designs implemented a steam release valve that kept the machine from exploding. By watching the valve rhythmically move up and down, Papin conceived of the idea of a piston and a cylinder engine. He did not, however, follow through with his design. Nevertheless, in 1697, based on Papin's designs, engineer Thomas Savery built the first engine, followed by Thomas Newcomen in 1712. Although these early engines were crude and inefficient, they attracted the attention of the leading scientists of the time.

The fundamental concepts of heat capacity and latent heat, which were necessary for the development of thermodynamics, were developed by Professor Joseph Black at the University of Glasgow, where James Watt was employed as an instrument maker. Black and Watt performed experiments together, but it was Watt who conceived the idea of the external condenser which resulted in a large increase in steam engine efficiency. Drawing on all the previous work led Sadi Carnot, the "father of thermodynamics", to publish *Reflections on the Motive Power of Fire* (1824), a discourse on heat, power, energy and engine efficiency. The paper outlined the basic energetic relations between the Carnot engine, the Carnot cycle, and motive power. It marked the start of thermodynamics as a modern science.

The first thermodynamic textbook was written in 1859 by William Rankine, originally trained as a physicist and a civil and mechanical engineering professor at the University of Glasgow. The first and second laws of thermodynamics emerged simultaneously in the 1850s, primarily out of the works of William Rankine, Rudolf Clausius, and William Thomson (Lord Kelvin).

The foundations of statistical thermodynamics were set out by physicists such as James Clerk Maxwell, Ludwig Boltzmann, Max Planck, Rudolf Clausius and J. Willard Gibbs.

During the years 1873-76 the American mathematical physicist Josiah Willard Gibbs published a series of three papers, the most famous being *On the Equilibrium of Heterogeneous Substances*, in which he showed how thermodynamic processes, including chemical reactions, could be graphically analysed, by studying the energy, entropy, volume, temperature and pressure of the thermodynamic system in such a manner, one can determine if a process would occur spontaneously. Also Pierre Duhem in the 19th century wrote about chemical thermodynamics. During the early 20th century, chemists such as Gilbert N. Lewis, Merle Randall, and E. A. Guggenheim applied the mathematical methods of Gibbs to the analysis of chemical processes.

Etymology

The etymology of *thermodynamics* has an intricate history. It was first spelled in a hyphenated form as an adjective (*thermo-dynamic*) and from 1854 to 1868 as the noun *thermo-dynamics* to represent the science of generalised heat engines.

The components of the word *thermodynamics* are derived from the Greek words *therme*, meaning heat, and *dynamis*, meaning power.

Pierre Perrot claims that the term *thermodynamics* was coined by James Joule in 1858 to designate the science of relations between heat and power. Joule, however, never used that term, but used instead the term *perfect thermo-dynamic engine* in reference to Thomson's 1849 phraseology.

By 1858, *thermo-dynamics*, as a functional term, was used in William Thomson's paper *An Account of Carnot's Theory of the Motive Power of Heat.*

Branches of Description

The study of thermodynamical systems has developed into several related branches, each using a different fundamental model as a theoretical or experimental basis, or applying the principles to varying types of systems.

Classical Thermodynamics

Classical thermodynamics is the description of the states (especially equilibrium states) and processes of thermodynamical systems, using macroscopic, empirical properties directly measurable in the laboratory. It is used to model exchanges of energy, work, heat, and matter, based on the laws of thermodynamics. The qualifier *classical* reflects the fact that it represents the descriptive level in terms of macroscopic empirical parameters that can be measured in the laboratory, that was the first level of understanding in the 19th century. A microscopic interpretation of these concepts was provided by the development of statistical thermodynamics.

Statistical Thermodynamics

Statistical thermodynamics, also called statistical mechanics, emerged with the development of atomic and molecular theories in the second half of the 19th century and early 20th century, supplementing thermodynamics with an interpretation of the microscopic interactions between individual particles or quantum-mechanical states. This field relates the microscopic properties of individual atoms and molecules to the macroscopic, bulk properties of materials that can be observed on the human scale, thereby explaining thermodynamics as a natural result of statistics, classical mechanics, and quantum theory at the microscopic level.

Chemical Thermodynamics

Chemical thermodynamics is the study of the interrelation of heat and work with chemical reactions or with physical changes of state within the confines of the laws of thermodynamics. Chemical thermodynamics involves not only laboratory measurements of various thermodynamic properties, but also the application of mathematical methods to the study of chemical questions and the *spontaneity* of processes.

The structure of chemical thermodynamics is based on the first two laws of thermodynamics. Starting from the first and second laws of thermodynamics, four equations called the "fundamental equations of Gibbs" can be derived. From these four, a multitude of equations, relating the thermodynamic properties of the thermodynamic system can be derived using relatively simple mathematics. This outlines the mathematical framework of chemical thermodynamics.

History

In 1865, the German physicist Rudolf Clausius, in his *Mechanical Theory of Heat*, suggested that the principles of thermochemistry, e.g. such as the heat evolved in combustion reactions, could be applied to the principles of thermodynamics. Building on the work of Clausius, between the years 1873-76 the American mathematical physicist Willard Gibbs published a series of three papers, the most famous one being the paper *On the Equilibrium of Heterogeneous Substances*. In these papers, Gibbs showed how the first two laws of thermodynamics could be measured graphically and mathematically to determine both the thermodynamic equilibrium of chemical reactions as well as their tendencies to occur or proceed. Gibbs' collection of papers provided the first unified body of thermodynamic theorems from the principles developed by others, such as Clausius and Sadi Carnot.

During the early 20th century, two major publications successfully applied the principles developed by Gibbs to chemical processes, and thus established the foundation of the science of chemical thermodynamics. The first was the 1923 textbook *Thermodynamics and the Free Energy of Chemical Substances* by Gilbert N. Lewis and Merle Randall. This book was responsible for supplanting the chemical affinity for the term free energy in the English-speaking world. The second was the 1933 book *Modern Thermodynamics by the methods of Willard Gibbs* written by E. A. Guggenheim. In this manner, Lewis, Randall, and Guggenheim are considered as the founders of modern chemical thermodynamics because of the major contribution of these two books in unifying the application of thermodynamics to chemistry.

Overview

The primary objective of chemical thermodynamics is the establishment of a criterion for the determination of the feasibility or spontaneity of a given transformation. In this manner, chemical thermodynamics is typically used to predict the energy exchanges that occur in the following processes:

1. Chemical reactions
2. Phase changes
3. The formation of solutions.

The following state functions are of primary concern in chemical thermodynamics:

- Internal energy (U)
- Enthalpy (H).

- Entropy (*S*)
- Gibbs free energy (*G*).

Most identities in chemical thermodynamics arise from application of the first and second laws of thermodynamics, particularly the law of conservation of energy, to these state functions.

The 3 laws of thermodynamics:

1. The energy of the universe is constant.
2. In any spontaneous process, there is always an increase in entropy of the universe
3. The entropy of a perfect crystal at 0 kelvins is zero.

Chemical Energy

Chemical energy is the potential of a chemical substance to undergo a transformation through a chemical reaction or to transform other chemical substances. Breaking or making of chemical bonds involves energy, which may be either absorbed or evolved from a chemical system.

Energy that can be released (or absorbed) because of a reaction between a set of chemical substances is equal to the difference between the energy content of the products and the reactants. This change in energy is called the change in internal energy of a chemical reaction. Where $\Delta U^{\circ}_{f\ \mathrm{reactants}}$ is the internal energy of formation of the reactant molecules that can be calculated from the bond energies of the various chemical bonds of the molecules under consideration and $\Delta U^{\circ}_{f\ \mathrm{products}}$ is the internal energy of formation of the product molecules. The internal energy change of a process is equal to the heat change if it is measured under conditions of constant volume, as in a closed rigid container such as a bomb calorimeter. However, under conditions of constant pressure, as in reactions in vessels open to the atmosphere, the measured heat change is not always equal to the internal energy change, because pressure-volume work also releases or absorbs energy. (The heat change at constant pressure is called the enthalpy change; in this case the enthalpy of formation).

Another useful term is the heat of combustion, which is the energy released due to a combustion reaction and often applied in the study of fuels. Food is similar to hydrocarbon fuel and carbohydrate fuels, and when it is oxidised, its caloric content is similar.

In chemical thermodynamics the term used for the chemical potential energy is chemical potential, and for chemical transformation an equation most often used is the Gibbs-Duhem equation.

Chemical Reactions

In most cases of interest in chemical thermodynamics there are internal degrees of freedom and processes, such as chemical reactions and phase transitions, which always create entropy unless they are at equilibrium, or are maintained at a "running equilibrium" through "quasi-static" changes by being coupled to constraining devices, such as pistons or electrodes, to deliver and receive external work. Even for homogeneous "bulk" materials, the free energy functions depend on the composition, as do all the extensive thermodynamic potentials, including the internal energy. If the quantities $\{N_i\}$, the number of chemical species, are omitted from the formulae, it is impossible to describe compositional changes.

Chemical Affinity

While this formulation is mathematically defensible, it is not particularly transparent since one does not simply add or remove molecules from a system. There is always a *process* involved in changing the composition; e.g., a chemical reaction (or many), or movement of molecules from one phase (liquid) to another (gas or solid). We should find a notation which does not seem to imply that the amounts of the components $(N_i\}$ can be changed independently. All real processes obey conservation of mass, and in addition, conservation of the numbers of atoms of each kind. Whatever molecules are transferred to or from should be considered part of the "system".

Consequently we introduce an explicit variable to represent the degree of advancement of a process, a progress variable ξ for the *extent of reaction* (Prigogine & Defay, p. 18; Prigogine, pp. 4–7; Guggenheim, p. 37.62), and to the use of the partial derivative $\partial G / \partial \xi$ (in place of the widely used "ΔG", since the quantity at issue is not a finite change). The result is an understandable expression for the dependence of dG on chemical reactions (or other processes). If there is just one reaction

$$(dG)_{T,P} = \left(\frac{\partial G}{\partial \xi}\right)_{T,P} d\xi.$$

If we introduce the *stoichiometric coefficient* for the *i-th* component in the reaction

$$\nu_i = \partial N_i / \partial \xi$$

which tells how many molecules of i are produced or consumed, we obtain an algebraic expression for the partial derivative

$$\left(\frac{\partial G}{\partial \xi}\right)_{T,P} = \sum_i \mu_i \nu_i = -\mathbb{A}$$

where, we introduce a concise and historical name for this quantity, the "affinity", symbolised by A, as introduced by Théophile de Donder in 1923. The minus sign comes from the fact the affinity was defined to represent the rule that spontaneous changes will ensue only when the change in the Gibbs free energy of the process is negative, meaning that the chemical species have a positive affinity for each other. The differential for G takes on a simple form which displays its dependence on compositional change

$$(dG)_{T,P} = -\mathbb{A}\, d\xi.$$

If there are a number of chemical reactions going on simultaneously, as is usually the case

$$(dG)_{T,P} = -\sum_k \mathbb{A}_k\, d\xi_k.$$

a set of reaction coordinates { ξ_j }, avoiding the notion that the amounts of the components (N_i } can be changed independently. The expressions above are equal to zero at thermodynamic equilibrium, while in the general case for real systems, they are negative because all chemical reactions proceeding at a finite rate produce entropy. This can be made even more explicit by introducing the reaction *rates* $d\xi_j/dt$. For each and every *physically independent process*

$$\mathbb{A}\,\dot{\xi} \le 0.$$

This is a remarkable result since the chemical potentials are intensive system variables, depending only on the local molecular milieu. They cannot "know" whether the temperature and pressure (or any other system variables) are going to be held constant over time. It is a purely local criterion and must hold regardless of any such constraints. Of course, it could have been obtained by taking partial derivatives of any of the other fundamental state functions, but nonetheless is a general criterion for (cT times) the entropy production from that spontaneous process; or at least any part of it that is not captured as external work.

We now relax the requirement of a homogeneous "bulk" system by letting the chemical potentials and the affinity apply to any locality in which a chemical reaction (or any other process) is occurring. By accounting for the entropy production due to irreversible processes, the inequality for dG is now replace by an equality

$$dG = -SdT + VdP - \sum_k \mathbb{A}_k \, d\xi_k + W'$$

or

$$dG_{T,P} = -\sum_k \mathbb{A}_k \, d\xi_k + W'.$$

Any decrease in the Gibbs function of a system is the upper limit for any isothermal, isobaric work that can be captured in the surroundings, or it may simply be dissipated, appearing as T times a corresponding increase in the entropy of the system and/or its surrounding. Or it may go partly towards doing external work and partly towards creating entropy. The important point is that the *extent of reaction* for a chemical reaction may be coupled to the displacement of some external mechanical or electrical quantity in such a way that one can advance only if the other one also does. The coupling may occasionally be *rigid*, but it is often flexible and variable.

Solutions

In solution chemistry and biochemistry, the Gibbs free energy decrease (eG/eω, in molar units, denoted cryptically by ΔG) is commonly used as a surrogate for (eT times) the entropy produced by spontaneous chemical reactions in situations where there is no work being done; or at least no "useful" work; i.e., other than perhaps some ± PdV. The assertion that all *spontaneous reactions have a negative ΔG* is merely a restatement of the fundamental thermodynamic relation, giving it the physical dimensions of energy and somewhat obscuring its significance in terms of entropy. When there is no useful work being done, it would be less misleading to use the Legendre transforms of the entropy appropriate for constant T, or for constant T and P, the Massieu functions eF/T and eG/T respectively.

Non Equilibrium

Generally the systems treated with the conventional chemical thermodynamics are either at equilibrium or near equilibrium. Ilya Prigogine developed the thermodynamic treatment of open systems that are far from equilibrium. In doing so he has discovered phenomena and structures of completely new and completely unexpected types. His generalised, non-linear and irreversible thermodynamics has found surprising applications in a wide variety of fields.

The non equilibrium thermodynamics has been applied for explaining how ordered structures e.g. the biological systems, can develop from disorder. Even if Onsager's relations are utilised, the classical principles of equilibrium in thermodynamics still show that linear systems close to equilibrium always develop into states of disorder which are stable to perturbations and cannot explain the occurrence of ordered structures.

Prigogine called these systems dissipative systems, because they are formed and maintained by the dissipative processes which take place because of the exchange of energy between the system and its environment and because they disappear if that exchange ceases. They may be said to live in symbiosis with their environment.

The method which Prigogine used to study the stability of the dissipative structures to perturbations is of very great general interest. It makes it possible to study the most varied problems, such as city traffic problems, the stability of insect communities, the development of ordered biological structures and the growth of cancer cells to mention but a few examples.

System Constraints

In this regard, it is crucial to understand the role of walls and other *constraints*, and the distinction between *independent* processes and *coupling*. Contrary to the clear implications of many reference sources, the previous analysis is not restricted to homogenous, isotropic bulk systems which can deliver only PdV work to the outside world, but applies even to the most structured systems. There are complex systems with many chemical "reactions" going on at the same time, some of which are really only parts of the same, overall process. An *independent* process is one that *could* proceed even if all others were unaccountably stopped in their tracks. Understanding this is perhaps a "thought experiment" in chemical kinetics, but actual examples exist.

A gas reaction which results in an increase in the number of molecules will lead to an increase in volume at constant external pressure. If it occurs inside a cylinder closed with a piston, the equilibrated reaction can proceed only by doing work against an external force on the piston. The extent variable for the reaction can increase only if the piston moves, and conversely, if the piston is pushed inward, the reaction is driven backwards.

Similarly, a redox reaction might occur in an electrochemical cell with the passage of current in wires connecting the electrodes. The half-cell reactions at the electrodes are constrained if no current is allowed to flow. The current might be dissipated as joule heating, or it might in turn run an electrical device like a motor doing mechanical work. An automobile lead-acid battery can be recharged, driving the chemical reaction backwards. In this case as well, the reaction is not an independent process. Some, perhaps most, of the Gibbs free energy of reaction may be delivered as external work.

The hydrolysis of ATP to ADP and phosphate can drive the force times distance work delivered by living muscles, and synthesis of ATP is in turn driven by a redox chain in mitochondria and chloroplasts, which involves the transport of ions across the membranes of these cellular organelles.

The coupling of processes here, and in the previous examples, is often not complete. Gas can leak slowly past a piston, just as it can slowly leak out of a rubber balloon. Some reaction may occur in a battery even if no external current is flowing. There is usually a coupling coefficient, which may depend on relative rates, which determines what percentage of the driving free energy is turned into external work, or captured as "chemical work"; a misnomer for the free energy of another chemical process.

Treatment of Equilibrium

Equilibrium thermodynamics studies transformations of matter and energy in systems as they approach equilibrium. The equilibrium means balance. In a thermodynamic equilibrium state there is no macroscopic flow and no macroscopic change is occurring or can be triggered; within the system, every microscopic process is balanced by its opposite; this is called the principle of detailed balance. A central aim in equilibrium thermodynamics is: given a system in a well-defined initial state, subject to accurately specified constraints, to calculate what the state of the system will be once it has reached equilibrium. A thermodynamic system is said to be homogeneous when all its locally defined intensive variables are spatially invariant. A simple system in thermodynamic equilibrium is homogeneous unless it is affected by a time-invariant externally imposed field of force, such as gravity, electricity, or magnetism.

Most systems found in nature are not in thermodynamic equilibrium, exactly considered; for they are changing or can be triggered to change over time, and are continuously and discontinuously subject to flux of matter and energy to and from other systems. Nevertheless, many systems in nature are nearly enough in thermodynamic equilibrium that for many purposes their behaviour can be nearly enough approximated by equilibrium calculations.

Non-Equilibrium Thermodynamics

Non-equilibrium thermodynamics is a branch of thermodynamics that deals with systems that are not in thermodynamic equilibrium. Most systems found in nature are not in thermodynamic equilibrium; for they are changing or can be triggered to change over time, and are continuously and discontinuously subject to flux of matter and energy to and from other systems and to chemical reactions. Non-equilibrium thermodynamics is concerned with transport processes and with the rates of chemical reactions. Many natural systems still today remain beyond the scope of currently known macroscopic thermodynamic methods.

The thermodynamic study of non-equilibrium systems requires more general concepts than are dealt with by equilibrium thermodynamics. One fundamental difference between equilibrium thermodynamics and non-equilibrium thermodynamics lies in the behaviour of inhomogeneous systems, which require for their study knowledge of rates of reaction which are not considered in equilibrium thermodynamics of homogeneous systems. This is discussed below. Another fundamental difference is the difficulty in defining entropy in macroscopic terms for systems not in thermodynamic equilibrium.

Overview

Non-equilibrium thermodynamics is a work in progress, not an established edifice.

Some concepts of particular importance for non-equilibrium thermodynamics include time rate of dissipation of energy (Rayleigh 1873, Onsager 1931, also), time rate of entropy production (Onsager 1931), thermodynamic fields, dissipative structure, and non-linear dynamical structure.

Of interest is the thermodynamic study of non-equilibrium steady states, in which entropy production and some flows are non-zero, but there is no time variation.

One initial approach to non-equilibrium thermodynamics might be called 'quasi-zero-dimensional'. There are other approaches to non-equilibrium thermodynamics, for example 'extended irreversible thermodynamics'.

Quasi-Radiationless Non-Equilibrium Thermodynamics of Matter in Laboratory Conditions

According to Wildt, current versions of non-equilibrium thermodynamics ignore radiant heat; they can do so because they refer to laboratory quantities of matter under laboratory conditions with temperatures well below the those of stars. At laboratory temperatures, in laboratory quantities of matter, thermal radiation is weak and can be practically nearly ignored. For example, atmospheric physics is concerned with large amounts of matter, occupying cubic kilometres, that are not necessarily within the scope of laboratory conditions.

'Quasi-Zero-Dimensional' Non-Equilibrium Thermodynamics

'Quasi-zero-dimensional' non-equilibrium thermodynamics demands certain simplifying assumptions, as follows. The assumptions have the effect of making the system effectively homogeneous, or well-mixed, or without an effective spatial structure, and without kinetic energy of bulk flow or diffusive flux. Even within the thought-frame of quasi-zero-dimensional non-equilibrium thermodynamics, care is needed in choosing the independent variables for systems.

In some writings, it is assumed that the intensive variables of equilibrium thermodynamics are sufficient as the independent variables for the task (such variables are considered to have no 'memory', and do not show hysteresis); in particular, local flow intensive variables are not admitted as independent variables; local flows are considered as dependent on static local intensive variables. (In other writings, local flow variables are considered; examples of the quasi-zero-dimensional approach with flow are in the thermoelectric phenomena known as the Seebeck and the Peltier effects, considered by Kelvin in the nineteenth century and by Onsager in the twentieth. These effects occur at metal junctions, which were originally effectively treated as two-dimensional surfaces, with no spatial volume, and no spatial variation.) Also it is assumed that the local entropy density is the same function of the other local intensive variables as in equilibrium; this is called the local thermodynamic equilibrium assumption.

In the quasi-zero-dimensional approach, it is usually assumed that there is no bulk flow of ponderable matter, so that kinetic energy does not need to be considered. Radiation is also ignored because it is transfer of energy between regions, which can be remote from one another. In a slight extension of quasi-zero dimensional non-equilibrium thermodynamics, spatial variation is allowed but it is assumed that the global entropy of the system can be found by simple spatial integration of the local entropy density; this means that spatial structure cannot contribute as it properly should to the global entropy assessment for the system. Also assumed in this slight extension is spatial and temporal continuity and even differentiability of locally defined intensive variables such as temperature and internal energy density. All of these are very stringent demands. Consequently, this approach can deal with only a very limited range of phenomena. This approach is nevertheless valuable as an introduction to non-equilibrium thermodynamics.

Basic Concepts

There are many examples of stationary non-equilibrium systems, some very simple, like a system confined between two thermostats at different temperatures or the ordinary Couette flow, a fluid enclosed between two flat walls moving in opposite directions and defining non-equilibrium conditions at the walls. Laser action is also a non-equilibrium process, but it depends on departure from local thermodynamic equilibrium and is thus beyond the scope of classical irreversible thermodynamics; here a strong temperature difference is maintained between two molecular degrees of freedom (with molecular laser, vibrational and rotational molecular motion), the requirement for two component 'temperatures' in the one small region of space, precluding local thermodynamic equilibrium, which demands that only one temperature be needed. Damping of acoustical perturbations or shock waves are non-stationary non-equilibrium processes. Driven complex fluids, turbulent systems and glasses are other examples of non-equilibrium systems.

The mechanics of macroscopic systems depends on a number of extensive quantities. It should be stressed that all systems are permanently interacting with their surroundings, thereby causing unavoidable fluctuations of extensive quantities. Equilibrium conditions of thermodynamic systems are related to the maximum property of the entropy. If the only extensive quantity that is allowed to fluctuate is the internal energy, all the other ones being kept strictly constant, the temperature of the system is measurable and meaningful. The system's properties are then most conveniently described using the thermodynamic potential Helmholtz free energy ($A = U - TS$), a Legendre transformation of the energy. If, next to fluctuations of the energy, the macroscopic dimensions (volume) of the system are left fluctuating, we use the Gibbs free energy ($G = U + PV - TS$), where the system's properties are determined both by the temperature and by the pressure. Non-equilibrium systems are much more complex and they may undergo fluctuations of more extensive quantities. The boundary conditions impose on them particular intensive variables, like temperature gradients or distorted collective motions (shear motions, vortices, etc.), often called thermodynamic forces. If free energies are very useful in equilibrium thermodynamics, it must be stressed that there is no general law defining stationary non-equilibrium properties of the energy as is the second law of thermodynamics for the entropy in equilibrium thermodynamics. That is why in such cases a more generalised Legendre transformation should be considered. This is the extended Massieu potential. By definition, the entropy (S) is a function of the collection of extensive quantities E_i. Each extensive quantity has a conjugate intensive variable I_i (a restricted definition of intensive variable is used here by comparison to the definition given in this link) so that:

$$I_i = \partial S / \partial E_i.$$

We then define the extended Massieu function as follows:

$$k_b M = S - \sum_i (I_i E_i),$$

where is Boltzmann's constant, whence

$$k_b\, dM = \sum_i (E_i\, dI_i).$$

The independent variables are the intensities.

Intensities are global values, valid for the system as a whole. When boundaries impose to the system different local conditions, (e.g. temperature differences), there are intensive variables representing the average value and others representing gradients or higher moments. The latter are the thermodynamic forces driving fluxes of extensive properties through the system.

It may be shown that the Legendre transformation changes the maximum condition of the entropy (valid at equilibrium) in a minimum condition of the extended Massieu function for stationary states, no matter whether at equilibrium or not.

Stationary States, Fluctuations, and Stability

In thermodynamics one is often interested in a stationary state of a process, allowing that the stationary state include the occurrence of unpredictable and experimentally unreproducible fluctuations in the state of the system. The fluctuations are due to the system's internal sub-processes and to exchange of matter or energy with the system's surroundings that create the constraints that define the process.

If the stationary state of the process is stable, then the unreproducible fluctuations involve local transient decreases of entropy. The reproducible response of the system is then to increase the entropy back to its maximum by irreversible processes: the fluctuation cannot be reproduced with a significant level of probability. Fluctuations about stable stationary states are extremely small except near critical points. The stable stationary state has a local maximum of entropy and is locally the most reproducible state of the system. There are theorems about the irreversible dissipation of fluctuations. Here 'local' means local with respect to the abstract space of thermodynamic coordinates of state of the system.

If the stationary state is unstable, then any fluctuation will almost surely trigger the virtually explosive departure of the system from the unstable stationary state. This can be accompanied by increased export of entropy.

Local Thermodynamic Equilibrium

The scope of present-day non-equilibrium thermodynamics does not cover all physical processes. A condition for the validity of many studies in non-equilibrium thermodynamics of matter is that they deal with what is known as *local thermodynamic equilibrium.*

Local Thermodynamic Equilibrium of Ponderable Matter

Local thermodynamic equilibrium of matter means that conceptually, for study and analysis, the system can be spatially and temporally divided into 'cells' or 'micro-phases' of small (infinitesimal) size, in which classical thermodynamical equilibrium conditions for matter are fulfilled to good approximation. These conditions are unfulfilled, for example, in very rarefied gases, in which molecular collisions are infrequent; and in the boundary layers of a star, where radiation is passing energy to space; and for interacting fermions at very low temperature, where dissipative processes become ineffective. When these 'cells' are defined, one admits that matter and energy may pass freely between contiguous 'cells', slowly enough to leave the 'cells' in their respective individual local thermodynamic equilibria with respect to intensive variables.

One can think here of two 'relaxation times' separated by order of magnitude. The longer relaxation time is of the order of magnitude of times taken for the macroscopic dynamical structure of the system to change. The shorter is of the order of magnitude of times taken for a single 'cell' to reach local thermodynamic equilibrium. If these two relaxation times are not well separated, then the classical non-equilibrium thermodynamical concept of local thermodynamic equilibrium loses its meaning. For example, in the atmosphere, the speed of sound is much greater than the wind speed; this favours the idea of local thermodynamic equilibrium of matter for atmospheric heat transfer studies at altitudes below about 60 km where sound propagates, but not above 100 km, where, because of the paucity of intermolecular collisions, sound does not propagate.

Milne's 1928 Definition of Local Thermodynamic Equilibrium in Terms of Radiative Equilibrium

Milne (1928), thinking about stars, gave a definition of 'local thermodynamic equilibrium' in terms of the thermal radiation of the matter in each small local 'cell'. He defined 'local thermodynamic equilibrium' in a 'cell' by requiring that it macroscopically absorb and spontaneously emit radiation as if it were in radiative equilibrium in a cavity at the temperature of the matter of the 'cell'. Then it strictly obeys Kirchhoff's law of equality of radiative emissivity and absorptivity, with a black body source function. The key to local thermodynamic equilibrium here is that the rate of collisions of ponderable matter particles such as molecules should far exceed the rates of creation and annihilation of photons.

Entropy in Evolving Systems

It is pointed out by W.T. Grandy Jr that entropy, though it may be defined for a non-equilibrium system, is when strictly considered, only a macroscopic quantity that refers to the whole system, and is not a dynamical variable and in general does not act as a local potential that describes local physical forces. Under special circumstances, however, one can metaphorically think as if the thermal variables behaved like local physical forces. The approximation that constitutes classical irreversible thermodynamics is built on this metaphoric thinking.

Flows and Forces

The fundamental relation of classical equilibrium thermodynamics

$$dS = \frac{1}{T}dU + \frac{p}{T}dV - \sum_{i=1}^{s}\frac{\mu_i}{T}dN_i$$

expresses the change in entropy dS of a system as a function of the intensive quantities temperature T, pressure p and i^{th} chemical potential μ_i and of the differentials of the extensive quantities energy U, volume V and i^{th} particle number N_i.

Following Onsager (1931,I), let us extend our considerations to thermodynamically non-equilibrium systems. As a basis, we need locally defined versions of the extensive macroscopic quantities U, V and N_i and of the intensive macroscopic quantities T, p and μ_i.

For classical non-equilibrium studies, we will consider some new locally defined intensive macroscopic variables. We can, under suitable conditions, derive these new variables by locally defining the gradients and flux densities of the basic locally defined macroscopic quantities.

Such locally defined gradients of intensive macroscopic variables are called 'thermodynamic forces'. They 'drive' flux densities, perhaps misleadingly often called 'fluxes', which are dual to the forces.

Establishing the relation between such forces and flux densities is a problem in statistical mechanics. Flux densities (J_i) may be coupled. Onsager reciprocal relations considers the stable near-steady thermodynamically non-equilibrium regime, which has dynamics linear in the forces and flux densities.

In stationary conditions, such forces and associated flux densities are by definition time invariant, as also are the system's locally defined entropy and rate of entropy production. Notably, according to Ilya Prigogine and others, when an open system is in conditions that allow it to reach a stable stationary thermodynamically non-equilibrium state, it organises itself so as to minimise total entropy production defined locally. This is considered further below.

One wants to take the analysis to the further stage of describing the behaviour of surface and volume integrals of non-stationary local quantities; these integrals are macroscopic fluxes and production rates. In general the dynamics of these integrals are not adequately described by linear equations, though in special cases they can be so described.

The Onsager Relations

Following Section III of Rayleigh (1873), Onsager (1931, I) showed that in the regime where both the flows are small and the thermodynamic forces vary slowly, there will be a linear relation between them, parametrised by a matrix of coefficients conventionally denoted L:

$$J_i = \sum_j L_{ij} \nabla I_j$$

The second law of thermodynamics requires that the matrix L be positive definite. Statistical mechanics considerations involving microscopic reversibility of dynamics imply that the matrix L is symmetric. This fact is called the *Onsager reciprocal relations.*

Speculated Thermodynamic Extremum Principles for Energy Dissipation and Entropy Production

Jou, Casas-Vazquez, Lebon (1993) note that classical non-equilibrium thermodynamics "has seen an extraordinary expansion since the second world war", and they refer to the Nobel prizes for work in the field awarded to Lars Onsager and Ilya Prigogine. Martyushev and Seleznev (2006) note the importance of entropy in the evolution of natural dynamical structures: "Great contribution has been done in this respect by two scientists, namely Clausius, ... , and Prigogine." Prigogine in his 1977 Nobel Lecture said: "... non-equilibrium may be a source of order. Irreversible processes may lead to a new type of dynamic states of matter which I have called "dissipative structures"." Glansdorff and Prigogine (1971) wrote on page xx: "Such 'symmetry breaking instabilities' are of special interest as they lead to a spontaneous 'self-organisation' of the system both from the point of view of its *space order* and its *function.*"

Analysing the Rayleigh-Bénard convection cell phenomenon, Chandrasekhar (1961) wrote "Instability occurs at the minimum temperature gradient at which a balance can be maintained between the kinetic energy dissipated by viscosity and the internal energy released by the buoyancy force." With a temperature gradient greater than the minimum, viscosity can dissipate kinetic energy as fast as it is released by convection due to buoyancy, and a steady state with convection is stable. The steady state with convection is often a pattern of macroscopically visible hexagonal cells with convection up or down in the middle or at the 'walls' of each cell, depending on the temperature dependence of the quantities; in the atmosphere under various conditions it seems that either is possible. (Some details are discussed by Lebon, Jou, and Casas-Vásquez (2008) on pages 143-158.) With a temperature gradient less than the minimum, viscosity and heat conduction are so effective that convection cannot keep going.

Glansdorff and Prigogine (1971) on page xv wrote "Dissipative structures have a quite different [from equilibrium structures] status: they are formed and maintained through the effect of exchange of energy and matter in non-equilibrium conditions." They were referring to the dissipation function of Rayleigh (1873) that was used also by Onsager (1931, I, 1931, II). On pages 78–80 of their book Glansdorff and Prigogine (1971) consider the stability of laminar flow that was pioneered by Helmholtz; they concluded that at a stable steady state of sufficiently slow laminar flow, the dissipation function was minimum.

These advances have led to proposals for various extremal principles for the "self-organised" régimes that are possible for systems governed by classical linear and non-linear non-equilibrium thermodynamical laws, with stable stationary régimes being particularly investigated. Convection introduces effects of momentum which appear as non-linearity in the dynamical equations. In the more restricted case of no convective motion, Prigogine wrote of "dissipative structures". Šilhavý (1997) offers the opinion that "... the extremum principles of [equilibrium] thermodynamics ... do not have any counterpart for [non-equilibrium] steady states (despite many claims in the literature)."

Prigogine's Proposed Theorem of Minimum Entropy Production

In 1945 Prigogine proposed a "Theorem of Minimum Entropy Production" which applies only to the linear regime near a stationary thermodynamically non-equilibrium state. The proof offered by Prigogine is open to serious criticism. A critical and unsupportive discussion of Prigogine's proposal is offered by Grandy (2008).

Speculated Principles of Maximum Entropy Production and Minimum Energy Dissipation

Onsager (1931, I) wrote: "Thus the vector field J of the heat flow is described by the condition that the rate of increase of entropy, less the dissipation function, be a maximum." Careful note needs to be taken of the opposite signs of the rate of entropy production and of the dissipation function, appearing in the left-hand side of Onsager's equation.

Although largely unnoticed at the time, Ziegler proposed an idea early with his work in the mechanics of plastics in 1961, and later in his book on thermomechanics revised in 1983, and in various papers (e.g., Ziegler (1987),). Ziegler never stated his principle as a universal law but he may have intuited this. He demonstrated his principle using vector space geometry based on an "orthogonality condition" which only worked in systems where the velocities were defined as a single vector or tensor, and thus, as he wrote at p. 347, was "impossible to test by means of macroscopic mechanical models", and was, as he pointed out, invalid in "compound systems where several elementary processes take place simultaneously".

In relation to the earth's atmospheric energy transport process, according to Tuck (2008), "On the macroscopic level, the way has been pioneered by a meteorologist (Paltridge 1975, 2001)." Initially Paltridge (1975) used the terminology "minimum entropy exchange", but after that, for example in Paltridge (1978), and in Paltridge (1979), he used the now current terminology "maximum entropy production" to describe the same thing. The logic of Paltridge's work is open to serious criticism. Paltridge (1978) cited Busse's (1967) fluid mechanical work concerning an extremum principle, but it seems apparent that Paltridge was misinterpreting his source. Nicolis and Nicolis (1980) discuss Paltridge's work, and they comment that the behaviour of the entropy production is far from simple and universal.

Sawada (1981), also in relation to the earth's atmospheric energy transport process, postulating a principle of largest amount of entropy increment per unit time, cites work in fluid mechanics by Malkus and Veronis (1958) as having "proven a principle of maximum heat current, which in turn is a maximum entropy production for a given boundary condition", but this inference is not logically valid. Again investigating planetary atmospheric dynamics, Shutts (1981) used an approach to the definition of entropy production, different from Paltridge's, to investigate a more abstract way to check the principle of maximum entropy production, and reported a good fit.

Prospects

At present, for this area of investigation, the prospects for useful extremal principles seem clouded at best. C. Nicolis (1999) concludes that one model of atmospheric dynamics has an attractor which is not a regime of maximum or minimum dissipation; she says this seems to rule out the existence of a global organising principle, and comments that this is to some extent disappointing; she also points to the difficulty of finding a thermodynamically consistent form of entropy production; in the present writer's opinion, there are few as expert in the theory of entropy production as Nicolis. Another top expert offers an extensive discussion of the possibilities for principles of extrema of entropy production and of dissipation of energy:

These views on the difficulty or impossibility of finding general global extremal principles are consistent with the views of Glansdorff and Prigogine (1971), and of Lebon, Jou and Casas-Vásquez (2008), and of Šilhavý (1997), noted.

Laws of Thermodynamics

The four laws of thermodynamics summarise the most important facts of thermodynamics. They define fundamental physical quantities, such as temperature, energy, and entropy, to describe thermodynamic systems and they describe the transfer of energy as heat and work in thermodynamic processes. Experimentally reproducible distinction between heat and work is at the heart of thermodynamics, and about processes in which this distinction cannot be made, thermodynamics has nothing to say.

Zeroth Law of Thermodynamics

The zeroth law of thermodynamics is a generalisation principle of thermal equilibrium among bodies, or thermodynamic systems, in contact. The zeroth law states that if two systems are in thermal equilibrium with a third system, they are also in thermal equilibrium with each other. Systems are said to be in thermal equilibrium if they do not actually exchange heat, and, if they are not already connected by a conductor of heat or pathway for exchange of thermal radiation, would not do so if they were so connected. The law implies that thermal equilibrium between systems is a transitive relation, which affords the definition of an empirical physical parameter, called temperature. The temperatures are equal for all systems in thermal equilibrium. The law permits the construction of a thermometer to measure this property.

Zeroth Law as Equivalence Relation

A system is said to be in thermal equilibrium when it experiences no net change in thermal energy. If *A*, *B*, and *C* are distinct thermodynamic systems, the zeroth law of thermodynamics can be expressed as:

"If *A* and *C* are each in thermal equilibrium with *B*, *A* is also in equilibrium with *C*."

This statement asserts that thermal equilibrium is a Euclidean relation between thermodynamic systems. If we also grant that all thermodynamic systems are in thermal equilibrium with themselves, then thermal equilibrium is also a reflexive relation. Relations that are both reflexive and Euclidean are equivalence relations. One consequence of this reasoning is that thermal equilibrium is a transitive relationship: If *A* is in thermal equilibrium with *B* and *B* is in thermal equilibrium with *C*, then *A* is in thermal equilibrium with *C*. Another consequence is that the equilibrium relationship is symmetric: If *A* is in thermal equilibrium with B, then B is in thermal equilibrium with A. Thus we may say that two systems are in thermal equilibrium with each other, or that they are in mutual equilibrium. Implicitly assuming both reflexivity and symmetry, the zeroth law is therefore often expressed as:

> *"If two systems are in thermal equilibrium with a third system, then they are in thermal equilibrium with each other"*

Again, implicitly assuming both reflexivity and symmetry, the zeroth law is occasionally expressed as the transitive relationship

> *"If* A *is in thermal equilibrium with* B *and if* B *is in thermal equilibrium with* C, *then* A *is in thermal equilibrium with* C."

Thermal Equilibrium between Many Systems

Many systems are said to be in equilibrium if the small, random exchanges (due to Brownian motion or photon emissions, for example) between them do not lead to a net change in the total energy summed over all systems. A simple example illustrates why the zeroth law is necessary to complete the equilibrium description.

Consider N systems in adiabatic isolation from the rest of the universe, i.e. no heat exchange is possible outside of these N systems, all of which have a constant volume and composition, and which can only exchange heat with one another.

The combined First and Second Laws relate the fluctuations in total energy, dU, to the temperature of the ith system, and the entropy fluctuation in the ith system, as follows:

$$dU = \sum_{i}^{N} T_i dS_i .$$

The adiabatic isolation of the system from the remaining universe requires that the total sum of the entropy fluctuations vanishes, or:

$$\sum_{i}^{N} dS_i = 0.$$

That is, entropy can only be exchanged between the N systems. This constraint can be used to rearrange the expression for the total energy fluctuation and obtain:

$$dU = \sum_{i}^{N} (T_i - T_j) dS_i ,$$

where T_j is the temperature of any system *j* we may choose to single out among the *N* systems. Finally, equilibrium requires the total fluctuation in energy to vanish, in which case:

$$\sum_{i}^{N}(T_i - T_j)dS_i = 0,$$

which can be thought of as the vanishing of the product of an anti-symmetric matrix $T_i - T_j$ and a vector of entropy fluctuations dS_i. In order for a non-trivial solution to exist,

$$dS_i \neq 0.$$

That is, the determinant of the matrix formed by $T_i - T_j$ must vanish for all choices of *N*. However, according to Jacobi's theorem, the determinant of a *N*x*N* anti-symmetric matrix is always zero if *N* is odd, although for *N* even we find that all of the entries must vanish, $T_i - T_j = 0$, in order to obtain a vanishing determinant. Hence $T_i - T_j$ at equilibrium. This non-intuitive result means that an odd number of systems are always in equilibrium regardless of their temperatures and entropy fluctuations, while equality of temperatures is only required between an even number of systems to achieve equilibrium in the presence of entropy fluctuations.

The zeroth law solves this odd vs. even paradox, because it can readily be used to reduce an odd-numbered system to an even number by considering any three of the *N* systems and eliminating one by application of its principle, and hence reduce the problem to even *N* which subsequently leads to the same equilibrium condition that we expect in every case, i.e., $T_i - T_j$. The same result applies to fluctuations in any extensive quantity, such as volume (yielding the equal pressure condition), or fluctuations in mass (leading to equality of chemical potentials). Hence the zeroth law has implications for a great deal more than temperature alone. In general, we see that the zeroth law breaks a certain kind of asymmetry present in the First and Second Laws.

Foundation of Temperature

The zeroth law establishes thermal equilibrium as an equivalence relationship. An equivalence relationship on a set (such as the set of thermally equilibrated systems) divides that set into a collection of distinct subsets ("disjoint subsets") where any member of the set is a member of one and only one such subset. In the case of the zeroth law, these subsets consist of systems which are in mutual equilibrium. This partitioning allows any member of the subset to be uniquely "tagged" with a label identifying the subset to which it belongs. Although the labelling may be quite arbitrary, temperature is just such a labelling process which uses the real number system for tagging. The zeroth law justifies the use of suitable thermodynamic systems as thermometers to provide such a labelling, which yield any number of possible empirical temperature scales, and justifies the use of the second law of thermodynamics to provide an absolute, or thermodynamic temperature scale. Such temperature scales bring additional continuity and ordering (i.e., "hot" and "cold") properties to the concept of temperature.

In the space of thermodynamic parameters, zones of constant temperature form a surface, that provides a natural order of nearby surfaces. One may therefore construct a global temperature function that provides a continuous ordering of states. The dimensionality of a surface of constant temperature is one less than the number of thermodynamic parameters, thus, for an ideal gas described with three thermodynamic parameters P, V and n, it is a two-dimensional surface.

For example, if two systems of ideal gases are in equilibrium, then $P_1V_1/N_1 = P_2V_2/N_2$ where P_i is the pressure in the ith system, V_i is the volume, and N_i is the amount (in moles, or simply the number of atoms) of gas.

The surface $PV/N = const$ defines surfaces of equal thermodynamic temperature, and one may label defining T so that $PV/N = RT$, where R is some constant. These systems can now be used as a thermometer to calibrate other systems. Such systems are known as "ideal gas thermometers".

History

Temperature has long been known as a quality of heat, for example, to Galileo and to Newton. Carnot took it as a presupposition for his work. Thermometers may be described as empirical or absolute. Absolute thermometers are calibrated numerically by the thermodynamic absolute temperature scale. It was not until the middle of the nineteenth century that absolute thermodynamic temperature was recognised, long after the recognition of empirical thermometry.

Empirical thermometry recognises hotness as a fundamental character of temperature and thermometers. Empirical thermometers are not in general necessarily in exact agreement with each other or with absolute thermometers as to their numerical scale readings, but to qualify as thermometers at all they must agree with absolute thermometers and with each other in the following way: given any two bodies isolated in their separate respective thermodynamic equilibrium states, all thermometers agree as to which of the two has the higher temperature, or that the two have equal temperatures. For any two empirical thermometers, this does not require that the relation between their numerical scale readings be linear, but it does require that relation to be strictly monotonic.

Truesdell reports that Rankine wrote in 1853:

Definition of equal temperatures.

Two portions of matter are said to have equal temperatures, when neither tends to communicate heat to the other.

Discussing the concept of temperature, James Clerk Maxwell in 1872 wrote: "If when two bodies are placed in thermal communication, one of the two bodies loses heat, and the other gains heat, that body which gives out heat is said to have a higher temperature than that which receives heat from it." He drew the corollary "If when two bodies are placed in thermal communication neither of them loses or gains heat, the two bodies are said to have equal temperatures or the same temperature. The two bodies are then said to be in thermal equilibrium."

Further, Maxwell stated, as the "Law of equal temperatures" the following triviality: "Bodies whose temperatures are equal to that of the same body have themselves equal temperatures". Maxwell then offered an argument that this statement was "not a truism". Later in the same text, Maxwell wrote: "Hence the result of the conduction and radiation of heat from one part of a system to another is to diminish the entropy of the system, or the energy, available as work, which can be obtained from the system." This statement was surrounded in Maxwell's text by several others like it that show that it was no slip of the pen. In the same textbook, Maxwell wrote that he was following Tait in re-defining the word entropy that had been introduced by Clausius. In contrast with what Maxwell wrote then, Tait had changed his mind by 1884 when in his text he accepted Clausius's original definition of entropy.

Subsequent writers made statements like Maxwell's. Tait in 1884 wrote "if A is at the same temperature as B and also at the same temperature as C — no transfer of heat takes place between B and C, whatever be these bodies." A similar statement was made by Max Planck in 1897, not labelled as a law but as an important proposition: "If a body, A, be in thermal equilibrium with two other bodies, B and

C, then B and C are in thermal equilibrium with one another." Planck repeated this important proposition in the seventh edition of his treatise in 1922.

The title "zeroth law of thermodynamics" began to appear in textbooks to refer to statements of this kind, though now stripped of their explicit reference to heat; their implicit dependence on the notion of heat could not be removed because they rely on the concept of thermal equilibrium which in turn relies on the concept of transfer of heat by conduction or radiation, the presence or absence of which must be empirically recognisable in order to make the concept of thermal equilibrium empirically recognisable. An early example is in the textbook of statistical thermodynamics of Fowler and Guggenheim (1939/1965). Their focus of interest in that book was homogeneous systems, which they termed 'assemblies'. They dealt with assemblies that were either completely homogeneous or that could be divided into homogeneous parts, called phases. For their macroscopic thermodynamic account of phenomena, they started by accepting, on empirical physical grounds, the presupposed notions of thermal insulation, thermal contact, and thermal equilibrium. They emphasised that these notions "can be defined without any reference to temperature". Moreover, they gave at this stage of their development of their theory no hint of the notion of heat transfer. They wrote:

> ...*we introduce the postulate:* If two assemblies are each in thermal equilibrium with a third assembly, they are in thermal equilibrium with each other.

They then proposed that "it may be shown to follow that the condition for thermal equilibrium between several assemblies is the equality of a certain single-valued function of the thermodynamics states of the assemblies, which may be called the temperature *t*, any one of the assemblies being used as a "thermometer" reading the temperature *t* on a suitable scale. This postulate of the "*Existence of temperature*" could with advantage be known as *the zeroth law of thermodynamics*". They did not there state any reason why such a function should have values in a scale consisting of a continuous succession of numbers or that it should have anything to do with heat. Their thinking was apparently conditioned by their asserted belief that "The most logically satisfactory formulation [of homogeneous system macroscopic thermodynamics] is undoubtedly that of Caratheodory". Though they thus apparently professed concern for logicality, they had no apparent compunction about immediately assuming, without apparent justification, that their postulated "temperatures" should exist on a numerical scale, provided for example by "the measured volume of a constant quantity of any chosen substance at constant pressure".

No worries about pesky but relevant physical realities such as the anomalous behaviour of water that concerned the nineteenth century thermodynamicists, because around 4 C it does not provide a valid empirical temperature. No mention that it would therefore be safer to refer to a permanent gas as a thermometric material. The approach of Fowler and Guggenheim is labelled "mechanical" by Bailyn, who contrasts it with the "thermodynamic" approach of Planck and the founders, who fully recognised the notion of heat transfer as an essential and fundamental presupposition to thermodynamics, without actually labelling it as a numbered law of thermodynamics. Since the time of Fowler and Guggenheim, who believed that the 1909 axiomatic formulation of Caratheodory was the most logically satisfactory, other influential axiomatic formulations of thermodynamics have appeared, some of which do not refer to a zeroth law of thermodynamics.

Sommerfeld in 1951 gave the title the "Zeroth Law" to the statement "Equality of temperature is a condition for thermal equilibrium between two systems or between two parts of a single system"; he wrote that this title followed the suggestion of Fowler, made when he was giving an account of a certain book. Sommerfeld's statement took the existence of temperature for granted, and used it to specify one of the characteristics of thermodynamic equilibrium. This is converse to many statements that are labelled as the zeroth law, which take thermal equilibrium for granted and use it to contribute to the concept of temperature. We may guess that Fowler had made his suggestion because the notion of temperature is in effect a presupposition of thermodynamics that earlier physicists had not felt needed explicit statement as a law of thermodynamics, and because the mood of his time, pursuing a "mechanical" axiomatic approach, wanted such an explicit statement.

Guggenheim in 1966 wrote "If two systems are both in thermal equilibrium with a third system then they are in thermal equilibrium with each other" as the *zeroth law* of thermodynamics, and followed it with the comment "In other words, systems in *thermal equilibrium* are said to have the *same temperature*." This ordinary language statement, although less precise than the statements of Planck and Tait mentioned above, conveys much of the essence of the zeroth law.

The statement of the zeroth law of thermodynamics by Serrin in 1977, though rather mathematically abstract, is more informative for empirical thermometry: "Zeroth Law - There exists a topological line *M* which serves as a coordinate manifold of material behaviour. The points *L* of the manifold *M* are called 'hotness levels', and *M* is called the 'universal hotness manifold'." To this information there needs to be added a sense of greater hotness; this sense can be had, independently of calorimetry, of thermodynamics, and of properties of particular materials, from Wien's displacement law of thermal radiation: the temperature of a bath of thermal radiation is proportional, by a universal constant, to the frequency of the maximum of its frequency spectrum; this frequency is always positive, but can have values that tend to zero.

First Law of Thermodynamics

The first law of thermodynamics is an expression of the principle of conservation of energy.

The law states that energy can be transformed, i.e. changed from one form to another, but cannot be created nor destroyed. It is usually formulated by stating that the change in the internal energy of a system is equal to the amount of heat supplied to the system, minus the amount of work performed by the system on its surroundings.

Original Statements

The first explicit statement of the first law of thermodynamics, by Rudolf Clausius in 1850, referred to cyclic thermodynamic processes.

"In all cases in which work is produced by the agency of heat, a quantity of heat is consumed which is proportional to the work done; and conversely, by the expenditure of an equal quantity of work an equal quantity of heat is produced."

Clausius stated the law also in another form, this time referring to the existence of a function of state of the system called the internal energy, and expressing himself in terms of a differential equation for the increments of a thermodynamic process. This equation may be translated into words as follows:

In a thermodynamic process, the increment in the internal energy of a system is equal to the difference between the increment of heat accumulated by the system and the increment of work done by it.

Description

The first law of thermodynamics was expressed in two ways by Clausius. One way referred to cyclic processes and the inputs and outputs of the system, but did not refer to increments in the internal state of the system. The other way referred to any incremental change in the internal state of the system, and did not expect the process to be cyclic. A cyclic process is one which can be repeated indefinitely often and still eventually leave the system in its original state.

In each repetition of a cyclic process, the work done by the system is proportional to the heat consumed by the system. In a cyclic process in which the system does work on its surroundings, it is necessary that some heat be taken in by the system and some be put out, and the difference is the heat consumed by the system in the process. The constant of proportionality is universal and independent of the system and was measured by Joule in 1845 and 1847.

In any incremental process, the change in the internal energy is considered due to a combination of heat added to the system and work done by the system. Taking dU as an infinitesimal (differential) change in internal energy, one writes

$$dU = \delta Q - \delta W$$

where δQ and δW are infinitesimal amounts of heat supplied to the system and work done by the system, respectively. Note that the minus sign in front of δW indicates that a positive amount of work done by the system leads to energy being lost from the system. (An alternate convention is to consider the work performed on the system by its surroundings. This leads to a change in sign of the work. This is the convention adopted by many modern textbooks of physical chemistry, such as those by Peter Atkins and Ira Levine, but many textbooks on physics define work as work done by the system.)

When a system expands in a quasistatic process, the work done on the environment is the product of pressure (P) and volume (V) change, i.e. PdV, whereas the work done on the system is $-PdV$. The change in internal energy of the system is:

$$dU = \delta Q - PdV.$$

Work and heat are expressions of actual physical processes which add or subtract energy, while U is a mathematical abstraction that keeps account of the exchanges of energy that befall the system. Thus the term heat for δQ means that amount of energy added as the result of heating, rather than referring to a particular form of energy. Likewise, work energy for δW means "that amount of energy lost as the result of work". Internal energy is a property of the system whereas work done and heat supplied are not. A significant result of this distinction is that a given internal energy change dU

can be achieved by, in principle, many combinations of heat and work.

The internal energy of a system is not uniquely defined. It is defined only up to an arbitrary additive constant of integration, which can be adjusted to give arbitrary reference zero levels. This non-uniqueness is in keeping with the abstract mathematical nature of the internal energy.

Evidence For the First Law of Thermodynamics

The first law of thermodynamics is induced from empirically observed evidence. The original discovery of the law was gradual over a period of perhaps half a century or more, and was mostly in terms of cyclic processes. The following is an account in terms of changes of state through compound processes that are not necessarily cyclic, but are composed of segments of two special kinds, adiabatic and isothermal diabatic.

Adiabatic Processes

It can be observed that, given a system in an initial state, if work is exerted on the system in an adiabatic (i.e. thermally insulated) way, the final state is the same for a given amount of work, irrespective of how this work is performed.

For instance, in Joule's experiment, the initial system is a tank of water with a paddle wheel inside. If we isolate thermally the tank and move the paddle wheel with a pulley and a weight we can relate the increase in temperature with the height descended by the mass. Now the system is returned to its initial state, isolated again, and the same amount of work is done on the tank using different devices (an electric motor, a chemical battery, a spring,...). In every case, the amount of work can be measured independently. The evidence shows that the final state of the water (in particular, its temperature) is the same in every case. It's irrelevant if the work is electrical, mechanical, chemical,... or if done suddenly or slowly, as long as it is performed in an adiabatic way.

This evidence leads to a statement of one aspect of the first law of thermodynamics

For all adiabatic processes between two specified states of a closed system, the net work done is the same regardless of the nature of the closed system and the details of the process.

This affirmation of path independence is one aspect of the meaning of the state function that is called internal energy, U. In an adiabatic process, adiabatic work takes the system from a reference state with internal energy $U(O)$ to an arbitrary eventual one with internal energy $U(A)$:

$$U(A) = U(O) + \Delta W^{\mathrm{ad}}_{O\to A}$$

where, following IUPAC convention we take as positive the work done on the system.

To go from a state A to a state B we can take a path that goes through the reference state, since the adiabatic work is independent of the path

$$\Delta W^{\mathrm{ad}}_{A\to B} = \Delta W^{\mathrm{ad}}_{A\to O} + \Delta W^{\mathrm{ad}}_{O\to B} = -\Delta W^{\mathrm{ad}}_{O\to A} + \Delta W^{\mathrm{ad}}_{O\to B} = -U(A) + U(B) = \Delta U$$

Isothermal Diabatic Processes

A complementary observable aspect of the first law is about heat transfer.

When the system does not evolve adiabatically, it is observed that the work exerted on the system does not coincide with the increase in its internal energy, which, being a state function, can be used for both adiabatic and non-adiabatic processes.

$$\Delta W^{non-ad} \neq \Delta U$$

The difference is due to the transfer of heat into the system, and the process is called diabatic. Heat transfer can be measured by calorimetry.

If the system is at constant temperature during the heat transfer, the transfer is called isothermal diabatic, and we may write ΔQ^{isoth}.

Putting the two complementary aspects together, adiabatic and isothermal diabatic, the inequality can be transformed into an equality as

$$\Delta W^{ad} + \Delta Q^{isoth} = \Delta U$$

This combined statement is the expression the first law of thermodynamics for finite processes composed of distinct adiabatic and isothermal diabatic segments.

In particular, if no work is exerted on a thermally isolated system we have

$$\Delta U = 0\,.$$

This is one aspect of the law of conservation of energy and can be stated:

The internal energy of an isolated system remains constant.

State Functional Formulation

The infinitesimal heat and work in the equations above are denoted by δ, rather than exact differentials denoted by *d*, because they do not describe the *state* of any system. The integral of an inexact differential depends upon the particular path taken through the space of thermodynamic parameters while the integral of an exact differential depends only upon the initial and final states. If the initial and final states are the same, then the integral of an inexact differential may or may not be zero, but the integral of an exact differential will always be zero. The path taken by a thermodynamic system through a chemical or physical change is known as a thermodynamic process.

An expression of the first law can be written in terms of exact differentials by realising that the work that a system does is, in case of a reversible process, equal to its pressure times the infinitesimal change in its volume. In other words $\delta w = PdV$ where P is pressure and V is volume. Also, for a reversible process, the total amount of heat added to a system can be expressed as $\delta Q = TdS$ where T is temperature and S is entropy. Therefore, for a reversible process:

$$dU = TdS - PdV.$$

Since U, S and V are thermodynamic functions of state, the above relation holds also for non-reversible changes. The above equation is known as the fundamental thermodynamic relation.

In the case where the number of particles in the system is not necessarily constant and may be of different types, the first law is written:

$$dU = \delta Q - \delta W + \sum_i \mu_i dN_i$$

where dN_i is the (small) number of type-i particles added to the system, and μ_i is the amount of energy added to the system when one type-i particle is added, where the energy of that particle is such that the volume and entropy of the system remains unchanged. μ_i is known as the chemical potential of the type-i particles in the system. The statement of the first law, using exact differentials is now:

$$dU = TdS - PdV + \sum_i \mu_i dN_i.$$

If the system has more external variables than just the volume that can change, the fundamental thermodynamic relation generalises to:

$$dU = TdS - \sum_i X_i dx_i + \sum_j \mu_j dN_j.$$

Here the X_i are the generalised forces corresponding to the external variables x_i.

A useful idea from mechanics is that the energy gained by a particle is equal to the force applied to the particle multiplied by the displacement of the particle while that force is applied. Now consider the first law without the heating term: $dU = d\ PdV$. The pressure P can be viewed as a force (and in fact has units of force per unit area) while dV is the displacement (with units of distance times area). We may say, with respect to this work term, that a pressure difference forces a transfer of volume, and that the product of the two (work) is the amount of energy transferred out of the system as a result of the process. If one were to make this term negative then this would be the work done on the system.

It is useful to view the TdS term in the same light: With respect to this heat term, a temperature difference forces a transfer of entropy, and the product of the two (heat) is the amount of energy transferred as a result of the process. Here, the temperature is known as a "generalised" force (rather than an actual mechanical force) and the entropy is a generalised displacement.

Similarly, a difference in chemical potential between groups of particles in the system forces a transfer of particles, and the corresponding product is the amount of energy transferred as a result of the process. For example, consider a system consisting of two phases: liquid water and water vapor. There is a generalised "force" of evaporation which drives water molecules out of the liquid. There is a generalised "force" of condensation which drives vapor molecules out of the vapor. Only when these two "forces" (or chemical potentials) are equal will there be equilibrium, and the net transfer will be zero.

The two thermodynamic parameters which form a generalised force-displacement pair are termed "conjugate variables". The two most familiar pairs are, of course, pressure-volume, and temperature-entropy.

Spatially Inhomogeneous Systems

Classical thermodynamics is focused on homogeneous systems, which might be regarded as 'zero-dimensional' in the sense that they have no spatial variation. But it is desired to study also systems with distinct internal motion and spatial inhomogeneity. For such systems, the principle of conservation of energy is expressed in terms not only of internal energy as defined for homogeneous systems, but also in terms of kinetic and potential energies. How the total energy of a system is allocated between these three more specific kinds of energy varies according to the purposes of different writers; this is because these components of energy are to some extent mathematical artefacts rather than actually measured physical quantities. If E denotes the total energy of a system, one may write

$$E = U + E_{kin} + E_{pot}$$

where E_{kin} and E_{pot} denote respectively the kinetic and potential energies internal to the system.

Potential energy can be exchanged with the surroundings of the system when the surroundings impose a force field, such as gravitational or electromagnetic, on the system.

The distinction between internal and kinetic energy is hard to make in the presence of turbulent motion within the system, as friction gradually dissipates macroscopic kinetic energy of localised bulk flow into molecular random motion of molecules that is classified as internal energy. The rate of dissipation by friction of kinetic energy of localised bulk flow into internal energy, whether in turbulent or in streamlined flow, is an important quantity in non-equilibrium thermodynamics. This is a serious difficulty for attempts to define entropy for time-varying spatially inhomogeneous systems.

History

The discovery of the first law of thermodymanics was by way of many tries and mistakes of investigation, over a period of about half a century. The first full statements of the law were made by Clausius in 1850 as noted above, and by Rankine also in 1850; Rankine's statement was perhaps not quite as clear and distinct as was Clausius'. A main aspect of the struggle was to deal with the previously proposed caloric theory of heat.

Germain Hess in 1840 stated a conservation law for so-called 'heat of reaction' for chemical reactions, but this was not explicitly concerned with the relation between energy exchanges by heat and work.

According to Truesdell (1980), Julius Robert von Mayer in 1841 made a statement that meant that "in a process at constant pressure, the heat used to produce expansion is universally interconvertible with work", but this is not a general statement of the first law.

Second Law of Thermodynamics

The second law of thermodynamics is an expression of the tendency that over time, differences in temperature, pressure, and chemical potential equilibrate in an isolated physical system. From the state of thermodynamic equilibrium, the law deduced the principle of the increase of entropy and explains the phenomenon of irreversibility in nature. The second law declares the impossibility of machines that generate usable energy from the abundant internal energy of nature by processes called *perpetual motion of the second kind.*

The second law may be expressed in many specific ways, but the first formulation is credited to the German scientist Rudolf Clausius. The law is usually stated in physical terms of impossible processes. In classical thermodynamics, the second law is a basic *postulate* applicable to any system involving measurable heat transfer, while in statistical thermodynamics, the second law is a *consequence* of unitarity in quantum theory. In classical thermodynamics, the second law defines the concept of thermodynamic entropy, while in statistical mechanics entropy is defined from information theory, known as the Shannon entropy.

Description

The first law of thermodynamics provides the basic definition of thermodynamic energy, also called internal energy, associated with all thermodynamic systems, but unknown in mechanics, and states the rule of conservation of energy in nature.

However, the concept of energy in the first law does not account for the observation that natural processes have a preferred direction of progress. For example, spontaneously, heat always flows to regions of lower temperature, never to regions of higher temperature without external work being performed on the system. The first law is completely symmetrical with respect to the initial and final states of an evolving system. The key concept for the explanation of this phenomenon through the second law of thermodynamics is the definition of a new physical property, the entropy.

A change in the entropy (S) of a system is the infinitesimal transfer of heat (Q) to a closed system driving a reversible process, divided by the equilibrium temperature (T) of the system.

$$dS = \frac{\delta Q}{T}$$

The entropy of an isolated system that is in equilibrium is constant and has reached its maximum value.

Empirical temperature and its scale is usually defined on the principles of thermodynamics equilibrium by the zeroth law of thermodynamics. However, based on the entropy, the second law permits a definition of the absolute, thermodynamic temperature, which has its null point at absolute zero.

The second law of thermodynamics may be expressed in many specific ways, the most prominent classical statements being the original statement by Rudolph Clausius (1850), the formulation by Lord Kelvin (1851), and the definition in axiomatic thermodynamics by Constantin Carathéodory (1909). These statements cast the law in general physical terms citing the impossibility of certain processes. They have been shown to be equivalent.

Clausius Statement

German scientist Rudolf Clausius is credited with the first formulation of the second law, now known as the *Clausius statement*:

No process is possible whose sole result is the transfer of heat from a body of lower temperature to a body of higher temperature.

Spontaneously, heat cannot flow from cold regions to hot regions without external work being performed on the system, which is evident from ordinary experience of refrigeration, for example. In a refrigerator, heat flows from cold to hot, but only when forced by an external agent, a compressor.

Kelvin Statement

Lord Kelvin expressed the second law in another form. The Kelvin statement expresses it as follows:

No process is possible in which the sole result is the absorption of heat from a reservoir and its complete conversion into work.

This means it is impossible to extract energy by heat from a high-temperature energy source and then convert all of the energy into work. At least some of the energy must be passed on to heat a low-temperature energy sink. Thus, a heat engine with 100% efficiency is thermodynamically impossible. This also means that it is impossible to build solar panels that generate electricity solely from the infrared band of the electromagnetic spectrum without consideration of the temperature on the other side of the panel (as is the case with conventional solar panels that operate in the visible spectrum).

Note that it is possible to convert heat completely into work, such as the isothermal expansion of ideal gas. However, such a process has an additional result. In the case of the isothermal expansion, the volume of the gas increases and never goes back without outside interference.

Principle of Carathéodory

Constantin Carathéodory formulated thermodynamics on a purely mathematical axiomatic foundation. His statement of the second law is known as the Principle of Carathéodory, which may be formulated as follows:

In every neighbourhood of any state S of an adiabatically isolated system there are states inaccessible from S.

With this formulation he described the concept of adiabatic accessibility for the first time and provided the foundation for a new subfield of classical thermodynamics, often called geometrical thermodynamics.

Equivalence of the Statements

Suppose there is an engine violating the Kelvin statement: i.e.,one that drains heat and converts it completely into work in a cyclic fashion without any other result. Now pair it with a reversed Carnot engine as shown by the graph. The net and sole effect of this newly created engine consisting of the two engines mentioned is transferring heat from $\Delta Q = Q\left(\frac{1}{\eta} - 1\right)$ the cooler reservoir to the hotter one, which violates the Clausius statement. Thus the Clausius statement implies the Kelvin statement. We can prove in a similar manner that the Kelvin statement implies the Clausius statement, or, in a word, the two are equivalent.

Corollaries

Perpetual Motion of the Second Kind: Prior to the establishment of the Second Law, many people who were interested in inventing a perpetual motion machine had tried to circumvent the restrictions of First Law of Thermodynamics by extracting the massive internal energy of the environment as the power of the machine. Such a machine is called a "perpetual motion machine of the second kind". The second law declared the impossibility of such machines.

Carnot Theorem

Carnot's theorem is a principle that limits the maximum efficiency for any possible engine. The efficiency solely depends on the temperature difference between the hot and cold thermal reservoirs. Carnot's theorem states:

- All irreversible heat engines between two heat reservoirs are less efficient than a Carnot engine operating between the same reservoirs.
- All reversible heat engines between two heat reservoirs are equally efficient with a Carnot engine operating between the same reservoirs.

Historically, the principle was based on the invalid caloric theory and preceded the establishment of the second law; however, it has since been recognised as a result of the second law.

Clausius Theorem

The Clausius theorem (1854) states that in a cyclic process

$$\oint \frac{\delta Q}{T} \leq 0.$$

The equality holds in the reversible case and the '<' is in the irreversible case. The reversible case is used to introduce the state function entropy. This is because in cyclic processes the variation of a state function is zero.

Thermodynamic Temperature

For an arbitrary heat engine, the efficiency is:

$$\eta = \frac{A}{q_H} = \frac{q_H - q_C}{q_H} = 1 - \frac{q_C}{q_H}$$

where A is the work done per cycle. Thus the efficiency depends only on q_C/q_H.

Carnot's theorem states that all reversible engines operating between the same heat reservoirs are equally efficient. Thus, any reversible heat engine operating between temperatures T_1 and T_2 must have the same efficiency, that is to say, the effiency is the function of temperatures only:

$$\frac{q_C}{q_H} = f(T_H, T_C).$$

In addition, a reversible heat engine operating between temperatures T_1 and T_3 must have the same efficiency as one consisting of two cycles, one between T_1 and another (intermediate) temperature T_2, and the second between T_2 and T_3. This can only be the case if

$$f(T_1, T_3) = \frac{q_3}{q_1} = \frac{q_2 q_3}{q_1 q_2} = f(T_1, T_2) f(T_2, T_3).$$

Now consider the case where T_1 is a fixed reference temperature: the temperature of the triple point of water. Then for any T_2 and T_3,

$$f(T_2,T_3) = \frac{f(T_1,T_3)}{f(T_1,T_2)} = \frac{273.16\cdot f(T_1,T_3)}{273.16\cdot f(T_1,T_2)}.$$

Therefore if thermodynamic temperature is defined by

$$T = 273.16\cdot f(T_1,T)$$

then the function *f*, viewed as a function of thermodynamic temperature, is simply

$$f(T_2,T_3) = \frac{T_3}{T_2},$$

and the reference temperature T_1 will have the value 273.16. (Of course any reference temperature and any positive numerical value could be used—the choice here corresponds to the Kelvin scale.)

Available Useful Work

An important and revealing idealised special case is to consider applying the Second Law to the scenario of an isolated system (called the total system or universe), made up of two parts: a sub-system of interest, and the sub-system's surroundings. These surroundings are imagined to be so large that they can be considered as an *unlimited* heat reservoir at temperature T_R and pressure P_R — so that no matter how much heat is transferred to (or from) the sub-system, the temperature of the surroundings will remain T_R; and no matter how much the volume of the sub-system expands (or contracts), the pressure of the surroundings will remain P_R.

Whatever changes to dS and dS_R occur in the entropies of the sub-system and the surroundings individually, according to the Second Law the entropy S_{tot} of the isolated total system must not decrease:

$$DS_{\text{tot}} = dS + dS_R \geq 0$$

According to the First Law of Thermodynamics, the change dU in the internal energy of the sub-system is the sum of the heat δq added to the sub-system, *less* any work δw done *by* the sub-system, *plus* any net chemical energy entering the sub-system $d\ d\mu_{iR}N_i$, so that:

$$DU = \delta q - \delta w + d(\sum \mu_{iR} N_i)$$

where μ_{iR} are the chemical potentials of chemical species in the external surroundings.

Now the heat leaving the reservoir and entering the sub-system is

$$\delta q = T_R(-dS_R) \leq T_R dS$$

where we have first used the definition of entropy in classical thermodynamics (alternatively, in statistical thermodynamics, the relation between entropy change, temperature and absorbed heat can be derived); and then the Second Law inequality from above.

It therefore follows that any net work δw done by the sub-system must obey

$$\delta w \leq -dU + T_R dS + \sum \mu_{iR} dN_i$$

It is useful to separate the work δw done by the subsystem into the *useful* work δw_u that can be done *by* the sub-system, over and beyond the work $p_R\ dV$ done merely by the sub-system expanding against the surrounding external pressure, giving the following relation for the useful work that can be done:

$$\delta w_u \leq -d(U - T_R S + p_R V - \sum \mu_{iR} N_i)$$

It is convenient to define the right-hand-side as the exact derivative of a thermodynamic potential, called the *availability* or exergy X of the subsystem,

$$X = U - T_R S + p_R V - \sum \mu_{iR} N_i$$

The Second Law therefore implies that for any process which can be considered as divided simply into a subsystem, and an unlimited temperature and pressure reservoir with which it is in contact,

$$DX + \delta w_u \leq 0$$

i.e. the change in the subsystem's exergy plus the useful work done *by* the subsystem (or, the change in the subsystem's exergy less any work, additional to that done by the pressure reservoir, done *on* the system) must be less than or equal to zero.

In sum, if a proper *infinite-reservoir-like* reference state is chosen as the system surroundings in the real world, then the Second Law predicts a decrease in X for an irreversible process and no change for a reversible process.

$$dS_{tot} \geq 0 \text{ Is equivalent to } dX + \delta W_u \leq 0$$

This expression together with the associated reference state permits a design engineer working at the macroscopic scale (above the thermodynamic limit) to utilise the Second Law without directly measuring or considering entropy change in a total isolated system. Those changes have already been considered by the assumption that the system under consideration can reach equilibrium with the reference state without altering the reference state. An efficiency for a process or collection of processes that compares it to the reversible ideal may also be found.

This approach to the Second Law is widely utilised in engineering practice, environmental accounting, systems ecology, and other disciplines to the formal energy created may not be supportive but it can be created from one form to another form ..it can be also called as conservation of energy.

History

The first theory of the conversion of heat into mechanical work is due to Nicolas Léonard Sadi Carnot in 1824. He was the first to realise correctly that the efficiency of this conversion depends on the difference of temperature between an engine and its environment.

Recognising the significance of James Prescott Joule's work on the conservation of energy, Rudolf Clausius was the first to formulate the second law during 1850, in this form: heat does not flow *spontaneously* from cold to hot bodies. While common knowledge now, this was contrary to the caloric theory of heat popular at the time, which considered heat as a fluid. From there he was able to infer the principle of Sadi Carnot and the definition of entropy (1865).

Established during the 19th century, the Kelvin-Planck statement of the Second Law says, "It is impossible for any device that operates on a cycle to receive heat from a single reservoir and produce a net amount of work." This was shown to be equivalent to the statement of Clausius.

The ergodic hypothesis is also important for the Boltzmann approach. It says that, over long periods of time, the time spent in some region of the phase space of microstates with the same energy is proportional to the volume of this region, i.e. that all accessible microstates are equally probable over a long period of time. Equivalently, it says that time average and average over the statistical ensemble are the same.

It has been shown that not only classical systems but also quantum mechanical ones tend to maximise their entropy over time. Thus the second law follows, given initial conditions with low entropy. More precisely, it has been shown that the local von Neumann entropy is at its maximum value with a very high probability. The result is valid for a large class of isolated quantum systems (e.g. a gas in a container). While the full system is pure and therefore does not have any entropy, the entanglement between gas and container gives rise to an increase of the local entropy of the gas. This result is one of the most important achievements of quantum thermodynamics.

Today, much effort in the field is attempting to understand why the initial conditions early in the universe were those of low entropy, as this is seen as the origin of the second law.

Informal Descriptions

The second law can be stated in various succinct ways, including:

- It is impossible to produce work in the surroundings using a cyclic process connected to a single heat reservoir (Kelvin, 1851).
- It is impossible to carry out a cyclic process using an engine connected to two heat reservoirs that will have as its only effect the transfer of a quantity of heat from the low-temperature reservoir to the high-temperature reservoir (Clausius, 1854).
- If thermodynamic work is to be done at a finite rate, free energy must be expended.

Mathematical Descriptions

In 1856, the German physicist Rudolf Clausius stated what he called the "second fundamental theorem in the mechanical theory of heat" in the following form:

$$\int \frac{\delta Q}{T} = -N$$

where Q is heat, T is temperature and N is the "equivalence-value" of all uncompensated transformations involved in a cyclical process. Later, in 1865, Clausius would come to define "equivalence-value" as entropy. On the heels of this definition, that same year, the most famous version of the second law was read in a presentation at the Philosophical Society of Zurich on April 24, in which, in the end of his presentation, Clausius concludes:

The entropy of the universe tends to a maximum.

This statement is the best-known phrasing of the second law. Moreover, owing to the general broadness of the terminology used here, e.g. universe, as well as lack of specific conditions, e.g. open, closed, or isolated, to which this statement applies, many people take this simple statement to mean that the second law of thermodynamics applies virtually to every subject imaginable. This, of course, is not true; this statement is only a simplified version of a more complex description.

In terms of time variation, the mathematical statement of the second law for an isolated system undergoing an arbitrary transformation is:

$$\frac{dS}{dt} \geq 0$$

where

S is the entropy and

t is time.

Statistical mechanics gives an explanation for the second law by postulating that a material is composed of atoms and molecules which are in constant motion. A particular set of positions and velocities for each particle in the system is called a microstate of the system and because of the constant motion, the system is constantly changing its microstate. Statistical mechanics postulates that, in equilibrium, each microstate that the system might be in is equally likely to occur, and when this assumption is made, it leads directly to the conclusion that the second law must hold in a statistical sense. That is, the second law will hold on average, with a statistical variation on the order of 1/"N where N is the number of particles in the system. For everyday (macroscopic) situations, the probability that the second law will be violated is practically zero. However, for systems with a small number of particles, thermodynamic parameters, including the entropy, may show significant statistical deviations from that predicted by the second law. Classical thermodynamic theory does not deal with these statistical variations.

Derivation from Statistical Mechanics

In statistical mechanics,the Second Law is not a postulate, rather it is a consequence of the fundamental postulate, also known as the equal prior probability postulate, so long as one is clear that simple probability arguments are applied only to the future, while for the past there are auxiliary sources of information which tell us that it was low entropy. The first part of the second law, which states that the entropy of a thermally isolated system can only increase is a trivial consequence of the equal prior probability postulate, if we restrict the notion of the entropy to systems in thermal equilibrium. The entropy of an isolated system in thermal equilibrium containing an amount of energy of E is:

$$S = k \log\left[\Omega(E)\right]$$

where Ω(E) is the number of quantum states in a small interval between E and $E + \delta E$. Here δE is a macroscopically small energy interval that is kept fixed. Strictly speaking this means that the entropy depends on the choice of δE. However, in the thermodynamic limit (i.e. in the limit of infinitely large system size), the specific entropy (entropy per unit volume or per unit mass) does not depend on δE.

Suppose we have an isolated system whose macroscopic state is specified by a number of variables. These macroscopic variables can, e.g., refer to the total volume, the positions of pistons in the system, etc. Then Ω will depend on the values of these variables. If a variable is not fixed, (e.g. we do not clamp a piston in a certain position), then because all the accessible states are equally likely in equilibrium, the free variable in equilibrium will be such that Ω is maximised as that is the most probable situation in equilibrium.

If the variable was initially fixed to some value then upon release and when the new equilibrium has been reached, the fact the variable will adjust itself so that Ω is maximised, implies that the entropy will have increased or it will have stayed the same (if the value at which the variable was fixed happened to be the equilibrium value).

The entropy of a system that is not in equilibrium can be defined as:

$$S = -k_B \sum_j P_j \log\left(P_j\right)$$

Here the P_j is the probabilities for the system to be found in the states labelled by the subscript j. In thermal equilibrium the probabilities for states inside the energy interval δE are all equal to 1 / Ω, and in that case the general definition coincides with the previous definition of S that applies to the case of thermal equilibrium.

Suppose we start from an equilibrium situation and we suddenly remove a constraint on a variable. Then right after we do this, there are a number Ω of accessible microstates, but equilibrium has not yet been reached, so the actual probabilities of the system being in some accessible state are not yet equal to the prior probability of $1 / \Omega$. We have already seen that in the final equilibrium state, the entropy will have increased or have stayed the same relative to the previous equilibrium state. Boltzmann's H-theorem, however, proves that the entropy will increase continuously as a function of time during the intermediate out of equilibrium state.

Derivation of the Entropy Change for Reversible Processes

The second part of the Second Law states that the entropy change of a system undergoing a reversible process is given by:

$$dS = \frac{\delta Q}{T}$$

where the temperature is defined as:

$$\frac{1}{kT} \equiv \beta \equiv \frac{d \log\left[\Omega(E)\right]}{dE}$$

See here for the justification for this definition. Suppose that the system has some external parameter, x, that can be changed. In general, the energy eigenstates of the system will depend on x. According to the adiabatic theorem of quantum mechanics, in the limit of an infinitely slow change of the system's Hamiltonian, the system will stay in the same energy eigenstate and thus change its energy according to the change in energy of the energy eigenstate it is in.

Derivation For Systems Described by the Canonical Ensemble

If a system is in thermal contact with a heat bath at some temperature T then, in equilibrium, the probability distribution over the energy eigenvalues are given by the canonical ensemble:

$$P_j = \frac{\exp\left(-\frac{E_j}{kT}\right)}{Z}$$

Here Z is a factor that normalises the sum of all the probabilities to 1, this function is known as the partition function. We now consider an infinitesimal reversible change in the temperature and in the external parameters on which the energy levels depend. It follows from the general formula for the entropy:

$$S = -k \sum_j P_j \log\left(P_j\right)$$

that

$$dS = -k\sum_j \log\left(P_j\right)dP_j$$

Inserting the formula for P_j for the canonical ensemble in here gives:

$$dS = \frac{1}{T}\sum_j E_j dP_j = \frac{1}{T}\sum_j d\left(E_j P_j\right) - \frac{1}{T}\sum_j P_j dE_j = \frac{dE + \delta W}{T} = \frac{\delta Q}{T}$$

General Derivation From Unitarity of Quantum Mechanics

The time development operator in quantum theory is unitary, because the Hamiltonian is hermitian. Consequently the transition probability matrix is doubly stochastic, which implies the Second Law of Thermodynamics. This derivation is quite general, based on the Shannon entropy, and does not require any assumptions beyond unitarity, which is universally accepted. It is a *consequence* of the irreversibility or singular nature of the general transition matrix.

Non-Equilibrium States

Statistically it is possible for a system to achieve moments of non-equilibrium. In such statistically unlikely events where hot particles "steal" the energy of cold particles enough that the cold side gets colder and the hot side gets hotter, for an instant. Such events have been observed at a small enough scale where the likelihood of such a thing happening is significant. The physics involved in such an event is described by the fluctuation theorem.

Controversies

Maxwell's Demon: James Clerk Maxwell imagined one container divided into two parts, *A* and *B*. Both parts are filled with the same gas at equal temperatures and placed next to each other. Observing the molecules on both sides, an imaginary demon guards a trapdoor between the two parts. When a faster-than-average molecule from *A* flies towards the trapdoor, the demon opens it, and the molecule will fly from *A* to *B*. The average speed of the molecules in *B* will have increased while in *A* they will have slowed down on average. Since average molecular speed corresponds to temperature, the temperature decreases in *A* and increases in *B*, contrary to the second law of thermodynamics.

One of the most famous responses to this question was suggested in 1929 by Leó Szilárd and later by Léon Brillouin. Szilárd pointed out that a real-life Maxwell's demon would need to have some means of measuring molecular speed, and that the act of acquiring information would require an expenditure of energy. But later exceptions were found.

Loschmidt's Paradox

Loschmidt's paradox, also known as the reversibility paradox, is the objection that it should not be possible to deduce an irreversible process from time-symmetric dynamics. This puts the time reversal symmetry of (almost) all known low-level fundamental physical processes at odds with any attempt to infer from them the second law of thermodynamics which describes the behaviour of macroscopic systems. Both of these are well-accepted principles in physics, with sound observational and theoretical support, yet they seem to be in conflict; hence the paradox.

One approach to handling Loschmidt's paradox is the fluctuation theorem, proved by Denis Evans and Debra Searles, which gives a numerical estimate of the probability that a system away from equilibrium will have a certain change in entropy over a certain amount of time. The theorem is proved with the exact time reversible dynamical equations of motion and the Axiom of Causality. The fluctuation theorem is proved utilising the fact that dynamics is time reversible. Quantitative predictions of this theorem have been confirmed in laboratory experiments at the Australian National University conducted by Edith M. Sevick et al. using optical tweezers apparatus.

Gibbs Paradox

In statistical mechanics, a simple derivation of the entropy of an ideal gas based on the Boltzmann distribution yields an expression for the entropy which is not extensive (is not proportional to the amount of gas in question). This leads to an apparent paradox known as the Gibbs paradox, allowing, for instance, the entropy of closed systems to decrease, violating the second law of thermodynamics.

The paradox is averted by recognising that the identity of the particles does not influence the entropy. In the conventional explanation, this is associated with an indistinguishability of the particles associated with quantum mechanics. However, a growing number of papers now take the perspective that it is merely the definition of entropy that is changed to ignore particle permutation (and thereby avert the paradox). The resulting equation for the entropy (of a classical ideal gas) is extensive, and is known as the Sackur-Tetrode equation.

Poincaré Recurrence Theorem

The Poincaré recurrence theorem states that certain systems will, after a sufficiently long time, return to a state very close to the initial state. The Poincaré recurrence time is the length of time elapsed until the recurrence, which is of the order of . The result applies to physical systems in which energy is conserved. The Recurrence theorem apparently contradicts the Second law of thermodynamics, which says that large dynamical systems evolve irreversibly towards the state with higher entropy, so that if one starts with a low-entropy state, the system will never return to it. There are many possible ways to resolve this paradox, but none of them is universally accepted. The most typical argument is that for thermodynamical systems like an ideal gas in a box, recurrence time is so large that for all practical purposes it is infinite.

Heat Death of the Universe

According to the second law, the entropy of any isolated system, such as the entire universe, never decreases. If the entropy of the universe has a maximum upper bound then when this bound is reached the universe has no thermodynamic free energy to sustain motion or life, that is, the heat death is reached.

Quotes

The law that entropy always increases holds, I think, the supreme position among the laws of Nature. If someone points out to you that your pet theory of the universe is in disagreement with Maxwell's equations — then so much the worse for Maxwell's equations. If it is found to be contradicted by observation — well, these experimentalists do bungle things sometimes. But if your theory is found to be against the second law of thermodynamics I can give you no hope; there is nothing for it but to collapse in deepest humiliation.

—Sir Arthur Stanley Eddington, The Nature of the Physical World (1927)

The tendency for entropy to increase in isolated systems is expressed in the second law of thermodynamics — perhaps the most pessimistic and amoral formulation in all human thought.

—Gregory Hill and Kerry Thornley, Principia Discordia (1965)

There are almost as many formulations of the second law as there have been discussions of it.

—Philosopher / Physicist P.W. Bridgman, (1941)

Third Law of Thermodynamics

The third law of thermodynamics is a statistical law of nature regarding entropy:

The entropy of a perfect crystal approaches zero as temperature approaches absolute zero.

For other materials, the residual entropy is not necessarily zero, although it is always zero for a perfect crystal in which there is only one possible ground state.

History

The third law was developed by the chemist Walther Nernst during the years 1906-1912, and is therefore often referred to as Nernst's theorem or Nernst's postulate. The third law of thermodynamics states that the entropy of a system at absolute zero is a well-defined constant. This is because a system at zero temperature exists in its ground state, so that its entropy is determined only by the degeneracy of the ground state. It means that *"it is impossible by any procedure, no matter how idealised, to reduce any system to the absolute zero of temperature in a finite number of operations"*.

An alternative version of the third law of thermodynamics as stated by Gilbert N. Lewis and Merle Randall in 1923:

If the entropy of each element in some (perfect) crystalline state be taken as zero at the absolute zero of temperature, every substance has a finite positive entropy; but at the absolute zero of temperature the entropy may become zero, and does so become in the case of perfect crystalline substances.

This version states not only ΔS will reach zero at 0 K, but S itself will also reach zero as long as the crystal has a ground state with only one configuration. Some crystals form defects which causes a residual entropy. This residual entropy disappears when the kinetic barriers to transitioning to one ground state are overcome.

With the development of statistical mechanics, the third law of thermodynamics (like the other laws) changed from a *fundamental* law (justified by experiments) to a *derived* law (derived from even more basic laws). The basic law from which it is primarily derived is the statistical-mechanics definition of entropy for a large system:

$$S = k_B \log \Omega$$

where S is entropy, k_B is the Boltzmann constant, and Ω is the number of microstates consistent with the macroscopic configuration.

Overview

In simple terms, the third law states that the entropy of a perfect crystal approaches zero as the absolute temperature approaches zero. This law provides an absolute reference point for the determination of entropy. The entropy determined relative to this point is the absolute entropy.

The entropy of a *perfect* crystal lattice as defined by Nernst's theorem is zero (provided that its ground state is unique, whereby ln(1)k = 0).

An example of a system which does not have a unique ground state is one containing half-integer spins, for which time-reversal symmetry gives two degenerate ground states (an entropy of ln(2) k_B, which is negligible on a macroscopic scale). Some crystalline systems exhibit geometrical frustration, where the structure of the crystal lattice prevents the emergence of a unique ground state. Ground-state helium (unless under pressure) remains liquid.

In addition, glasses and solid solutions retain large entropy at 0K, because they are large collections of nearly degenerate states, in which they become trapped out of equilibrium. Another example of a solid with many nearly-degenerate ground states, trapped out of equilibrium, is ice Ih, which has "proton disorder".

For the third law to apply strictly, the magnetic moments of a perfectly ordered crystal must themselves be perfectly ordered; indeed, from an entropic perspective, this can be considered to be part of the definition of "perfect crystal". Only ferromagnetic, anti-ferromagnetic, and diamagnetic materials can satisfy this condition. Materials that remain paramagnetic at 0K, by contrast, may have many nearly-degenerate ground states (for example, in a spin glass), or may retain dynamic disorder (a spin liquid).

9

Thermodynamic System

A thermodynamic system is a precisely defined macroscopic region of the universe, often called a physical system, that is studied using the principles of thermodynamics.

All space in the universe outside the thermodynamic system is known as the *surroundings*, the *environment*, or a *reservoir*. A system is separated from its surroundings by a *boundary* which may be notional or real, but which by convention delimits a finite volume. Exchanges of work, heat, or matter between the system and the surroundings may take place across this boundary. Thermodynamic systems are often classified by specifying the nature of the exchanges that are allowed to occur across its boundary.

A thermodynamic system is characterised and defined by a set of thermodynamic parameters associated with the system. The parameters are experimentally measurable macroscopic properties, such as volume, pressure, temperature, electric field, and others.

The set of thermodynamic parameters necessary to uniquely define a system is called the thermodynamic state of a system. The state of a system is expressed as a functional relationship, the equation of state, between its parameters. A system is in thermodynamic equilibrium when the state of the system does not change with time.

Originally, in 1824, Sadi Carnot described a thermodynamic system as the working substance under study.

Overview

Thermodynamics describes the physics of matter using the concept of the *thermodynamic system*, a region of the universe that is under study. All quantities, such as pressure or mechanical work, in an equation refer to the system unless labelled otherwise. As thermodynamics is fundamentally concerned with the flow and balance of energy and matter, systems are distinguished depending on the kinds of interaction they undergo and the types of energy they exchange with the surrounding environment.

Isolated systems are completely isolated from their environment. They do not exchange heat, work or matter with their environment. An example of an isolated system is a completely insulated rigid container, such as a completely insulated gas cylinder. *Closed systems* are able to exchange energy (heat and work) but not matter with their environment. A greenhouse is an example of a closed system exchanging heat but not work with its environment. Whether a system exchanges heat, work or both is usually thought of as a property of its boundary. *Open systems* may exchange any form of energy as well as matter with their environment. A boundary allowing matter exchange is called permeable. The ocean would be an example of an open system.

In practice, a system can never be absolutely isolated from its environment, because there is always at least some slight coupling, such as gravitational attraction. In analysing a system in steady-state, the energy into the system is equal to the energy leaving the system.

An example system is the system of hot liquid water and solid table salt in a sealed, insulated test tube held in a vacuum (the surroundings). The test tube constantly loses heat in the form of black-body radiation, but the heat loss progresses very slowly. If there is another process going on in the test tube, for example the dissolution of the salt crystals, it will probably occur so quickly that any heat lost to the test tube during that time can be neglected. Thermodynamics in general does not measure time, but it does sometimes accept limitations on the time frame of a process.

History

The first to develop the concept of a thermodynamic system was the French physicist Sadi Carnot whose 1824 *Reflections on the Motive Power of Fire* studied what he called the *working substance*, e.g., typically a body of water vapor, in steam engines, in regards to the system's ability to do work when heat is applied to it. The working substance could be put in contact with either a heat reservoir (a boiler), a cold reservoir (a stream of cold water), or a piston (to which the working body could do work by pushing on it). In 1850, the

German physicist Rudolf Clausius generalised this picture to include the concept of the surroundings, and began referring to the system as a "working body." In his 1850 manuscript *On the Motive Power of Fire,* Clausius wrote:

> *"With every change of volume (to the working body) a certain amount work must be done by the gas or upon it, since by its expansion it overcomes an external pressure, and since its compression can be brought about only by an exertion of external pressure. To this excess of work done by the gas or upon it there must correspond, by our principle, a proportional excess of heat consumed or produced, and the gas cannot give up to the "surrounding medium" the same amount of heat as it receives."*

Carnot heat engine shows the original piston-and-cylinder diagram used by Carnot in discussing his ideal engine; below, we see the Carnot engine as is typically modelled in current use:

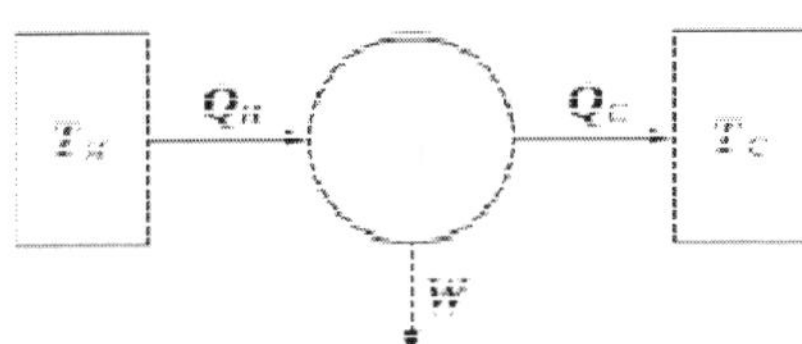

Carnot engine diagram (modern) - where heat flows from a high temperature T_H furnace through the fluid of the "working body" (working substance) and into the cold sink T_C, thus forcing the working substance to do mechanical work W on the surroundings, via cycles of contractions and expansions.

The "working body" (system), a term introduced by Clausius in 1850, can be any fluid or vapor body through which heat Q can be introduced or transmitted through to produce work. In 1824, Sadi Carnot, in his famous paper *Reflections on the Motive Power of Fire,* had postulated that the fluid body could be any substance capable of expansion, such as vapor of water, vapor of alcohol, vapor of mercury, a permanent gas, or air, etc.

Although, in these early years, engines came in a number of configurations, typically Q_H was supplied by a boiler, wherein water was boiled over a furnace; Q_C was typically a stream of cold flowing water in the form of a condenser located on a separate part of the engine. The output work W here is the movement of the piston as it is used to turn a crank-arm, which was then typically used to turn a pulley so to lift water out of flooded salt mines. Carnot defined work as "weight lifted through a height."

Boundary

A system boundary is a real or imaginary two-dimensional closed surface that encloses or demarcates the volume or region that a thermodynamic system occupies, across which quantities such as heat, mass, or work can flow. In short, a thermodynamic boundary is a geometrical division between a system and its surroundings. Topologically, it is usually considered to be nearly or piecewise smoothly homeomorphic with a two-sphere, because a system is usually considered to be simply connected.

Boundaries can also be fixed (e.g. a constant volume reactor) or moveable (e.g. a piston). For example, in an engine, a fixed boundary means the piston is locked at its position; as such, a constant volume process occurs. In that same engine, a moveable boundary allows the piston to move in and out. Boundaries may be real or imaginary. For closed systems, boundaries are real while for open system boundaries are often imaginary. For theoretical purposes, a boundary may be declared to be adiabatic, isothermal, diathermal, insulating, permeable, or semipermeable, but actual physical materials that provide such idealised properties are not always readily available.

Anything that passes across the boundary that effects a change in the internal energy needs to be accounted for in the energy balance equation. The volume can be the region surrounding a single atom resonating energy, such as Max Planck defined in 1900; it can be a body of steam or air in a steam engine, such as Sadi Carnot defined in 1824; it can be the body of a tropical cyclone, such as Kerry Emanuel theorised in 1986 in the field of atmospheric thermodynamics; it could also be just one nuclide (i.e. a system of quarks) as hypothesised in quantum thermodynamics.

Surroundings

The system is the part of the universe being studied, while the *surroundings* is the remainder of the universe that lies outside the boundaries of the system. It is also known as the *environment*, and the *reservoir*. Depending on the type of system, it may interact with the system by exchanging mass, energy (including heat and work), momentum, electric charge, or other conserved properties. The environment is ignored in analysis of the system, except in regards to these interactions.

Open System

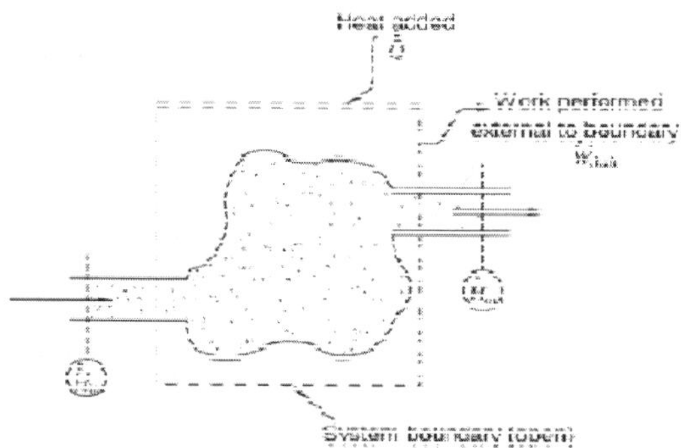

Figure: *During steady, continuous operation, an energy balance applied to an open system equates shaft work performed by the system to heat added plus net enthalpy added.*

In open systems, matter may flow in and out of the system boundaries. The first law of thermodynamics for open systems states: *the increase in the internal energy of a system is equal to the amount of energy added to the system by matter flowing in and by heating, minus the amount lost by matter flowing out and in the form of work done by the system.* The first law for open systems is given by:

$$\mathrm{d}U = \mathrm{d}U_{in} + \delta Q - \mathrm{d}U_{out} - \delta W$$

where U_{in} is the average internal energy entering the system and U_{out} is the average internal energy leaving the system

The region of space enclosed by open system boundaries is usually called a control volume, and it may or may not correspond to physical walls. If we choose the shape of the control volume such that all flow in or out occurs perpendicular to its surface, then the flow of matter into the system performs work as if it were a piston of fluid pushing mass into the system, and the system performs work on the flow of matter out as if it were driving a piston of fluid. There are then two types of work performed: *flow work* described above which is performed on the fluid (this is also often called *PV work*) and *shaft work* which may be performed on some mechanical device. These two types of work are expressed in the equation:

$$\delta W = \mathrm{d}(P_{out}V_{out}) - \mathrm{d}(P_{in}V_{in}) + \delta W_{shaft}$$

Substitution into the equation above for the control volume *cv* yields:

$$\mathrm{d}U_{cv} = \mathrm{d}U_{in} + \mathrm{d}(P_{in}V_{in}) - \mathrm{d}U_{out} - \mathrm{d}(P_{out}V_{out}) + \delta Q - \delta W_{shaft}$$

The definition of enthalpy, H, permits us to use this thermodynamic potential to account for both internal energy and PV work in fluids for open systems:

$$\mathrm{d}U_{cv} = \mathrm{d}H_{in} - \mathrm{d}H_{out} + \delta Q - \delta W_{shaft}$$

During steady-state operation of a device (*see turbine, pump, and engine*), any system property within the control volume is independent of time. Therefore, the internal energy of the system enclosed by the control volume remains constant, which implies that $\mathrm{d}U_{cv}$ in the expression above may be set equal to zero. This yields a useful expression for the power generation or requirement for these devices in the absence of chemical reactions:

$$\frac{\delta W_{shaft}}{\mathrm{d}t} = \frac{\mathrm{d}H_{in}}{\mathrm{d}t} - \frac{\mathrm{d}H_{out}}{\mathrm{d}t} + \frac{\delta Q}{\mathrm{d}t}$$

This expression is described by the diagram above.

Closed System

In a closed system, no mass may be transferred in or out of the system boundaries. The system will always contain the same amount of matter, but heat and work can be exchanged across the boundary of the system. Whether a system can exchange heat, work, or both is dependent on the property of its boundary.

- Adiabatic boundary – not allowing any heat exchange
- Rigid boundary – not allowing exchange of work

One example is fluid being compressed by a piston in a cylinder. Another example of a closed system is a bomb calorimeter, a type of constant-volume calorimeter used in measuring the heat of combustion of a particular reaction. Electrical energy travels across the boundary to produce a spark between the electrodes and initiates combustion. Heat transfer occurs across the boundary after combustion but no mass transfer takes place either way.

Beginning with the first law of thermodynamics for an open system, this is expressed as:

$$\mathrm{d}U = Q - W + m_i(h + \frac{1}{2}v^2 + gz)_i - m_e(h + \frac{1}{2}v^2 + gz)_e$$

where U is internal energy, Q is heat transfer, W is work, and since no mass is transferred in or out of the system, both expressions involving mass flow, , zeroes, and the first law of thermodynamics for a closed system is derived. The first law of thermodynamics for a closed system states that the amount of internal energy within the system equals the difference between the amount of heat added to or extracted from the system and the work done by or to the system. The first law for closed systems is stated by:

$$dU = \delta Q - \delta W$$

where U is the average internal energy within the system, Q is the heat added to or extracted from the system and W is the work done by or to the system.

Substituting the amount of work needed to accomplish a reversible process, which is stated by:

$$\delta W = P dV$$

where P is the measured pressure and V is the volume, and the heat required to accomplish a reversible process stated by the second law of thermodynamics, the universal principle of entropy, stated by:

$$\delta Q = T dS$$

where T is the absolute temperature and S is the entropy of the system, derives the fundamental thermodynamic relationship used to compute changes in internal energy, which is expressed as:

$$\delta U = T dS - P dV$$

For a simple system, with only one type of particle (atom or molecule), a closed system amounts to a constant number of particles. However, for systems which are undergoing a chemical reaction, there may be all sorts of molecules being generated and destroyed by the reaction process. In this case, the fact that the system is closed is expressed by stating that the total number of each elemental atom is conserved, no matter what kind of molecule it may be a part of. Mathematically:

$$\sum_{j=1}^{m} a_{ij} N_j = b_i^0$$

where N_j is the number of j-type molecules, a_{ij} is the number of atoms of element i in molecule j and b_i^0 is the total number of atoms of element i in the system, which remains constant, since the system is closed. There will be one such equation for each different element in the system.

Isolated System

An isolated system is more restrictive than a closed system as it does not interact with its surroundings in any way. Mass and energy remains constant within the system, and no energy or mass transfer takes place across the boundary. As time passes in an isolated system, internal differences in the system tend to even out and pressures and temperatures tend to equalise, as do density differences. A system in which all equalising processes have gone practically to completion is considered to be in a state of thermodynamic equilibrium.

Truly isolated physical systems do not exist in reality (except perhaps for the universe as a whole), because, for example, there is always gravity between a system with mass and masses elsewhere. However, real systems may behave nearly as an isolated system for finite (possibly very long) times. The concept of an isolated system can serve as a useful model approximating many real-world situations. It is an acceptable idealisation used in constructing mathematical models of certain natural phenomena.

In the attempt to justify the postulate of entropy increase in the second law of thermodynamics, Boltzmann's H-theorem used equations which assumed a system (for example, a gas) was isolated. That is all the mechanical degrees of freedom could be specified, treating the walls simply as mirror boundary conditions. This inevitably led to Loschmidt's paradox. However, if the stochastic behaviour of the molecules in actual walls is considered, along with the randomising effect of the ambient, background thermal radiation, Boltzmann's assumption of molecular chaos can be justified.

The second law of thermodynamics is only true for isolated systems. It states that the entropy of an isolated system not in equilibrium will tend to increase over time, approaching maximum value at equilibrium. Overall, in an isolated system, the available energy can never increase, and it complement, entropy, can never decrease. A closed system's entropy can decrease.

It is important to note that isolated systems are not equivalent to closed systems. Closed systems cannot exchange matter with the surroundings, but can exchange energy. Isolated systems can exchange neither matter nor energy with their surroundings, and as such are only theoretical and do not exist in reality (except, possibly, the entire universe). It is worth noting that 'closed system' is often used in thermodynamics discussions when 'isolated system' would be correct - i.e. there is an assumption that energy does not enter or leave the system.

Systems in Equilibrium

At thermodynamic equilibrium, a system's properties are, by definition, unchanging in time. Systems in equilibrium are much simpler and easier to understand than systems which are not in equilibrium. Often, when analysing a thermodynamic process, it can be assumed that each intermediate state in the process is at equilibrium. This will also considerably simplify the analysis.

In isolated systems it is consistently observed that as time goes on internal rearrangements diminish and stable conditions are approached. Pressures and temperatures tend to equalise, and matter arranges itself into one or a few relatively homogeneous phases. A system in which all processes of change have gone practically to completion is considered to be in a state of thermodynamic equilibrium. The thermodynamic properties of a system in equilibrium are unchanging in time. Equilibrium system states are much easier to describe in a deterministic manner than non-equilibrium states.

In thermodynamic processes, large departures from equilibrium during intermediate steps are associated with increases in entropy and increases in the production of heat rather than useful work. It can be shown that for a process to be reversible, each step in the process must be reversible. For a step in a process to be reversible, the system must be in equilibrium throughout the step. That ideal cannot be accomplished in practice because no step can be taken without perturbing the system from equilibrium, but the ideal can be approached by making changes slowly.

Thermodynamic Instruments

A thermodynamic instrument is any device which facilitates the quantitative measurement of thermodynamic systems. In order for a thermodynamic parameter to be truly defined, a technique for its measurement must be specified. For example, the ultimate definition of temperature is "what a thermometer reads". The question follows - what is a thermometer?

There are two types of thermodynamic instruments, the meter and the reservoir. A thermodynamic meter is any device which measures any parameter of a thermodynamic system. A thermodynamic reservoir is a system which is so large that it does not appreciably alter its state parameters when brought into contact with the test system.

Overview

Two general complementary tools are the meter and the reservoir. It is important that these two types of instruments are distinct. A meter does not perform its task accurately if it behaves like a reservoir of the state variable it is trying to measure. If, for example, a thermometer, were to act as a temperature reservoir it would alter the temperature of the system being measured, and the reading would be incorrect. Ideal meters have no effect on the state variables of the system they are measuring.

Thermodynamic Meters

A meter is a thermodynamic system which displays some aspect of its thermodynamic state to the observer. The nature of its contact with the system it is measuring can be controlled, and it is sufficiently small that it does not appreciably affect the state of the system being measured. The theoretical thermometer described below is just such a meter.

In some cases, the thermodynamic parameter is actually defined in terms of an idealised measuring instrument. For example, the zeroth law of thermodynamics states that if two bodies are in thermal equilibrium with a third body, they are also in thermal equilibrium with each other. This principle, as noted by James Maxwell in 1872, asserts that it is possible to measure temperature. An idealised thermometer is a sample of an ideal gas at constant pressure. From the ideal gas law, the volume of such a sample can be used as an indicator of temperature; in this manner it defines temperature. Although pressure is defined mechanically, a pressure-measuring device called a barometer may also be constructed from a sample of an ideal gas held at a constant temperature. A calorimeter is a device which is used to measure and define the internal energy of a system.

Some common thermodynamic meters are:

- Thermometer - a device which measures temperature as described above
- Barometer - a device which measures pressure. An ideal gas barometer may be constructed by mechanically connecting an ideal gas to the system being measured, while thermally insulating it. The volume will then measure pressure, by the ideal gas equation $P=NkT/V$.
- Calorimeter - a device which measures the heat energy added to a system. A simple calorimeter is simply a thermometer connected to a thermally isolated system.

Thermodynamic Reservoirs

A reservoir is a thermodynamic system which controls the state of a system, usually by "imposing" itself upon the system being controlled. This means that the nature of its contact with the system can be controlled. A reservoir is so large that its thermodynamic state is not appreciably affected by the state of the system being controlled. The term "atmospheric pressure" in the below description of a theoretical thermometer is essentially a "pressure reservoir" which imposes atmospheric pressure upon the thermometer.

Some common reservoirs are:

- Pressure reservoir - by far the most common pressure reservoir is the Earth's atmosphere.
- Temperature reservoir - A large quantity of water at its triple point forms an effective temperature reservoir.

Theory

Let's assume that we understand mechanics well enough to understand and measure volume, area, mass, and force. These may be combined to understand the concept of pressure, which is force per unit area and density, which is mass per unit volume. It has been experimentally determined that, at low enough pressures and densities, all gases behave as ideal gases. The behaviour of an ideal gas is given by the ideal gas law:

$$PV - NkT$$

where P is pressure, V is volume, N is the number of particles (total mass divided by mass per particle), k is Boltzmann's constant, and T is temperature. In fact, this equation is more than a phenomenological equation, it gives an operational, or experimental, definition of temperature. A thermometer is a tool that measures temperature - a primitive thermometer would simply be a small container of an ideal gas, that was allowed to expand against atmospheric pressure. If we bring it into thermal contact with the system whose temperature we wish to measure, wait until it equilibrates, and then measure the volume of the thermometer, we will be able to calculate the temperature of the system in question via $T=PV/Nk$. Hopefully, the thermometer will be small enough that it does not appreciably alter the temperature of the system it is measuring, and also that the atmospheric pressure is not affected by the expansion of the thermometer.

The ideal gas thermometer can be defined more precisely by saying it is a system containing an ideal gas, which is thermally connected to the system it is measuring, while being dynamically and materially insulated from it. It is simultaneously dynamically connected to an external pressure reservoir, from which it is materially and thermally insulated. Other thermometers (e.g. mercury thermometers, which display the volume of mercury to the observer) may now be constructed, and calibrated against the ideal gas thermometer.

Conjugate Variables (Thermodynamics)

In thermodynamics, the internal energy of a system is expressed in terms of pairs of conjugate variables such as temperature/entropy or pressure/volume. In fact all thermodynamic potentials are expressed in terms of conjugate pairs.

For a mechanical system, a small increment of energy is the product of a force times a small displacement. A similar situation exists in thermodynamics. An increment in the energy of a thermodynamic system can be expressed as the sum of the products of certain generalised "forces" which, when imbalanced, cause certain generalised "displacements", and the product of the two is the energy transferred as a result. These forces and their associated displacements are called conjugate variables. The thermodynamic force is always an intensive variable and the displacement is always an extensive variable, yielding an extensive energy transfer. The intensive (force) variable is the derivative of the internal energy with respect to the extensive (displacement) variable, while all other extensive variables are held constant.

The thermodynamic square can be used as a tool to recall and derive some of the thermodynamic potentials based on conjugate variables.

Overview

Just as a small increment of energy in a mechanical system is the product of a force times a small displacement, so an increment in the energy of a thermodynamic system can be expressed as the sum of the products of certain generalised "forces" which, when unbalanced, cause certain generalised "displacements" to occur, with their product being the energy transferred as a result. These forces and their associated displacements are called conjugate variables. For example, consider the pV conjugate pair. The pressure P acts as a generalised force: Pressure differences force a change in volume dV, and their product is the energy lost by the system due to work. Here pressure is the driving force, volume is the associated displacement, and the two form a pair of conjugate variables. In a similar way, temperature differences drive changes in entropy, and their product is the energy transferred by heat transfer. The thermodynamic force is always an intensive variable and the displacement is always an extensive variable, yielding an extensive energy. The intensive (force) variable is the derivative of the (extensive) internal energy with respect to the extensive (displacement) variable, with all other extensive variables held constant.

The theory of thermodynamic potentials is not complete until we consider the number of particles in a system as a variable on par with the other extensive quantities such as volume and entropy. The number of particles is, like volume and entropy, the displacement variable in a conjugate pair. The generalised force component of this pair is the chemical potential. The chemical potential may be thought of as a force which, when imbalanced, pushes an exchange of particles, either with the surroundings, or between phases inside the system.

In cases where there are a mixture of chemicals and phases, this is a useful concept. For example if a container holds liquid water and water vapor, there will be a chemical potential (which is negative) for the liquid which pushes the water molecules into the vapor (evaporation) and a chemical potential for the vapor, pushing vapor molecules into the liquid (condensation). Only when these "forces" equilibrate, and the chemical potentials of each phase is equal, is equilibrium obtained.

The most commonly considered conjugate thermodynamic variables are (with corresponding SI units):

Thermal parameters:

- Temperature: T (K)
- Entropy: S (J K^{-1})

Mechanical parameters:

- Pressure: P (Pa= J m^{-3})
- Volume: V (m^3 = J Pa^{-1})

or, more generally,

- Stress: σ_{ij} (Pa= J m^{-3})
- Volume × Strain: $V \times \varepsilon_{ij}$ (m^3 = J Pa^{-1})

$U[\mu_1,\mu_2] = U - \mu_1 N_1 - \mu_2 N_2$ *and so on. If there are* D *dimensions to the thermodynamic space, then there are* 2^D *unique thermodynamic potentials. For the most simple case, a single phase ideal gas, there will be three dimensions, yielding eight thermodynamic potentials.*

In statistical mechanics, the relationship between the Helmholtz free energy and the partition function is fundamental, and is used to calculate the thermodynamic properties of matters.

Axiomatics

Most accounts of thermodynamics presuppose the law of conservation of mass, sometimes with, and sometimes without, explicit mention. Particular attention is paid to the law in accounts of non-equilibrium thermodynamics. One statement of this law is "The total mass of [a closed] system remains constant." Another statement of it is "In a chemical reaction, matter is neither created nor destroyed." Implied in this is that matter and energy are not considered to be interconverted in such accounts. The full generality of the law of conservation of energy is thus not used in such accounts.

In 1909, Constantin Carathéodory presented a purely mathematical axiomatic formulation, a description often referred to as geometrical thermodynamics, and sometimes said to take the "mechanical approach" to thermodynamics. The Carathéodory formulation is restricted to equilibrium thermodynamics and does not attempt to deal with non-equilibrium thermodynamics, forces that act at a distance on the system, or surface tension effects.

Moreover, Carathéodory's formulation does not deal with materials like water near 4C, which have a density extremum as a function of temperature at constant pressure. Carathéodory used the law of conservation of energy as an axiom from which, along with the contents of the zeroth law, and some other assumptions including his own version of the second law, he derived the first law of thermodynamics. Consequently one might also describe Carathëodory's work as lying in the field of energetics, which is broader than thermodynamics. Carathéodory presupposed the law of conservation of mass without explicit mention of it.

Since the time of Carathëodory, other influential axiomatic formulations of thermodynamics have appeared, which like Carathéodory's, use their own respective axioms, different from the usual statements of the four laws, to derive the four usually stated laws.

Many axiomatic developments assume the existence of states of thermodynamic equilibrium and of states of thermal equilibrium.

States of thermodynamic equilibrium of compound systems allow their component simple systems to exchange heat and matter and to do work on each other on their way to overall joint equilibrium. Thermal equilibrium allows them only to exchange heat. The physical properties of glass depend on its history of being heated and cooled and, strictly speaking, glass is not in thermodynamic equilibrium.

According to Herbert Callen's widely cited 1985 text on thermodynamics: "An essential prerequisite for the measurability of energy is the existence of walls that do not permit transfer of energy in the form of heat." According to Werner Heisenberg's mature and careful examination of the basic concepts of physics, the theory of heat has a self-standing place.

From the viewpoint of the axiomatist, there are several different ways of thinking about heat, temperature, and the second law of thermodynamics. The Clausius way rests on the empirical fact that heat is conducted always down, never up, a temperature gradient. The Kelvin way is to assert the empirical fact that conversion of heat into work by cyclic processes is never perfectly efficient. A more mathematical way is to assert the existence of a function of state called the entropy which tells whether a hypothesised process will occur spontanteously in nature. A more abstract way is that of Carathéodory that in effect asserts the irreversibility of some adiabatic processes. For these different ways, there are respective corresponding different ways of viewing heat and temperature.

The Clausius-Kelvin-Planck way This way prefers ideas close to the empirical origins of thermodynamics. It presupposes transfer of energy as heat, and empirical temperature as a scalar function of state. According to Gislason and Craig (2005): "Most thermodynamic data come from calorimetry..." According to Kondepudi (2008): "Calorimetry is widely used in present day laboratories." In this approach, what is often currently called the zeroth law of thermodynamics is deduced as a simple consequence of the presupposition of the nature of heat and empirical temperature, but it is not named as a numbered law of thermodynamics. Planck attributed this point of view to Clausius, Kelvin, and Maxwell. Planck wrote that he thought that no proof of the second law of thermodynamics could ever work that was not based on the impossibility of a perpetual motion machine of the second kind. In that treatise, Planck makes no mention of the 1909 Carathéodory way, which was well-known by 1922. Planck for himself chose a version of what is just above called the Kelvin way. The development by

Truesdell and Bharatha (1977) is so constructed that it can deal naturally with cases like that of water near 4C.

The way that assumes the existence of entropy as a function of state This way also presupposes transfer of energy as heat, and it presupposes the usually stated form of the zeroth law of thermodynamics, and from these two it deduces the existence of empirical temperature. Then from the existence of entropy it deduces the existence of absolute thermodynamic temperature.

The Carathéodory way This way presupposes that the state of a simple one-phase system is fully specifiable by just one more state variable than the known exhaustive list of mechanical variables of state. It does not explicitly name empirical temperature, but speaks of the one-dimensional "non-deformation coordinate". This satisfies the definition of an empirical temperature, that lies on a one-dimensional manifold. The Carathéodory way needs to assume moreover that the one-dimensional manifold has a definite sense, which determines the direction of irreversible adiabatic process, which is effectively assuming that heat is conducted from hot to cold. This way presupposes the often currently stated version of the zeroth law, but does not actually name it as one of its axioms.

Technical Drawing

Technical drawing, also known as drafting or draughting, is the act and discipline of composing plans that visually communicate how something functions or is to be constructed.

A *drafter*, *draftsperson*, or *draughtsman* is a person who makes a drawing (technical or otherwise). A professional drafter who makes technical drawings is sometimes called a *drafting technician*.

People who communicate with technical drawings may use technical standards that define practical symbols, perspectives, units of measurement, notation systems, visual styles, or layout conventions. These enable a drafter to communicate more concisely by using a commonly-understood convention. Together, such conventions constitute a visual language, and help to ensure that the drawing is unambiguous and relatively easy to understand.

This need for unambiguous communication in the preparation of a functional document distinguishes technical drawing from the expressive drawing of the visual arts. Artistic drawings are subjectively interpreted; their meanings are multiply determined. Technical drawings are understood to have one intended meaning.

Methods

Sketching: A sketch is a quickly executed freehand drawing that is not intended as a finished work. In general, sketching is a quick way to record an idea for later use. Architect's sketches primarily serve as a way to try out different ideas and establish a composition before undertaking a more finished work, especially when the finished work is expensive and time consuming.

These sketches, like metaphors, are used by architects as a mean of communication in aiding design collaboration. This tool helps architects to abstract attributes of hypothetical provisional design solutions and summarise their complex patterns, hereby enhancing the design process.

Manual or by Instrument

The basic drafting procedure is to place a piece of paper (or other material) on a smooth surface with right-angle corners and straight sides—typically a drawing board. A sliding straightedge known as a *T-square* is then placed on one of the sides, allowing it to be slid across the side of the table, and over the surface of the paper.

"Parallel lines" can be drawn simply by moving the T-square and running a pencil or technical pen along the T-square's edge, but more typically the T-square is used as a tool to hold other devices such as set squares or triangles. In this case the drafter places one or more triangles of known angles on the T-square—which is itself at right angles to the edge of the table—and can then draw lines at any chosen angle to others on the page. Modern drafting tables (which have by now largely been replaced by CAD workstations) come equipped with a drafting machine that is supported on both sides of the table to slide over a large piece of paper. Because it is secured on both sides, lines drawn along the edge are guaranteed to be parallel.

In addition, the drafter uses several tools to draw curves and circles. Primary among these are the compasses, used for drawing simple arcs and circles, and the French curve, typically a piece of plastic with complex curves on it. A spline is a rubber coated articulated metal that can be manually bent to most curves.

Drafting templates assist the drafter with creating recurring objects in a drawing without having to reproduce the object from scratch every time. This is especially useful when using common symbols; i.e. in the context of stagecraft, a lighting designer will

Latrobe, Benjamin Henry: *The Engineering Drawings of Benjamin Henry Latrobe,* Yale University Press, NY, 1980.

Lincoln, B., Gomes, K.J., Braden, J.F.: *Mechanical Fastening of Plastics - An Engineering Handbook,* Marcel Dekker, New York, 1984.

Melvin W. Brown: *Seals and Sealing Handbook,* Elsevier Advanced Technology, Oxford, 1995.

Myer Kutz: *Mechanical Engineers' Handbook,* John Wiley & Sons, New York, 1998.

P.K. Mallick: *Composites Engineering Handbook,* M. Dekker, New York, 1997.

Parsons, William Barclay: *Engineers and Engineering in the Renaissance,* Williams & Wilkins, Baltimore, 1939.

Reader, W. J.: *A History of the Institution of Electrical Engineers, 1871-1971,* Peregrinus, London, 1987.

Ronald K. Jurgen: *Automotive Electronics Handbook,* McGraw-Hill, New York, 1999.

S.T. Peters: *Handbook of Composites,* Chapman & Hall, London, 1998.

Satish G. Kandlikar, Masahiro Shoji, Vijay K. Dhir: *Handbook of Phase Change: Boiling and Condensation,* Taylor & Francis, Philadelphia, PA, 1999.

Sinclair, Bruce: *A Centennial History of the American Society of Mechanical Engineers, 1880-1980,* University of Toronto Press, Toronto, 1980.

Teng H. Hsu: *Structural Engineering & Applied Mechanics Data Handbook,* Gulf Pub. Co., Houston, 1988.

Tres, P.A.: *Designing Plastic Parts for Assembly,* Hanser Publishers, Munich, 1998.

Tyler G. Hicks: *Handbook of Mechanical Engineering Calculations,* McGraw-Hill, New York, 1998.

Vincenti, Walter: *What Engineers Know and How They Know It,* Johns Hopkins University Press, Baltimore, 1997.

Wilbert F. Stoecker: *Industrial Refrigeration Handbook,* McGraw-Hill, New York, 1998.

Index

❑❑❑